T0226382

# Introduction to Transport Phenomena Modeling

Gianpaolo Ruocco

# Introduction to Transport Phenomena Modeling

## A Multiphysics, General Equation-Based Approach

 Springer

Gianpaolo Ruocco
College of Engineering
University of Basilicata
Potenza
Italy

ISBN 978-3-319-88323-6      ISBN 978-3-319-66822-2   (eBook)
https://doi.org/10.1007/978-3-319-66822-2

This Springer imprint is published by the registered company Springer Nature Switzerland AG
The registered company address is: Gewerbestrasse 11, 6330 Cham, Switzerland

*Alle voci lontane dei bambini che giocano in giardino,*
*allo sguardo notturno che scruta le stelle,*[1]
*a Beatrice e Roberto Maria.*
*Alla mia Musa.*
FLVCTVAT·NEC·MERGITUR[2]

---

[1]To children voices from a distant playground, to the glance at night searching for stars,…

[2]My motto since I was a boy: in Latin, "The ship is tossed by the waves, nevertheless it does not sink." At the beginning, I was unaware that this was also an old motto of the city of Paris, France: based on what occurred to me so far, it turned out quite fit.

# Preface

**Motivation** The vast subject of Transport Phenomena is gaining importance in today's technology perspective, as *momentum, heat, and mass transport* are found in many engineering processes, for both industrial and environmental frameworks.

Environmental flows in porous media, wind patterns around city buildings along with traffic pollutants, the thermal response of water-saturated substrates when subject to drying by microwaves or ultrasounds, the formation of harmful compounds in food during cooking, the colonization of bacteria or the progression of cancer: the applications of Transport Phenomena in our lives or in professional activities are endless.

In this book, an introductory approach is provided which presents these topics through an orderly *application of partial differential equations* (PDEs), leading to the exploitation of *mathematical fields* through their analytical or numerical solution. *Modeling* of PDEs-driven phenomena has its inner workings that need to be recognized and understood. The final goal is that one can achieve a process *virtualization*, i.e., the replica of what we observe in the process reality and around us.

**Background** In the past 10 years of my teaching and research activity, I have been dealing with many cases of thermal processing to multiphase substrates. I soon realized that intertwined occurrences were frequent in these cases: to complete the study on momentum and heat transport, one must include mass transport for these mechanisms were always linked each other due to the inherent phase- or composition-changing. In this book, a generalized procedure to model this interdependence is presented.

**Opportunity** Over the years, the computing technology has considerably evolved and so had the engineering analysis and modeling that exploits it. Nowadays, with the development of robust and efficient numerical techniques, the computation of interdependent Transport Phenomena is a valid tool to realistic process description. In a broad sense, the simultaneous existence of more mechanisms at once in the same process, cutting across the fields of physics, chemistry, mechanics, and biotechnology,[3] can be called **multiphysics**.

Engineering graduates face more frequently with modeling challenges. Many academic programs offer specific courses but they are usually restricted to chemical engineering curricula. Nevertheless, the **unifying language of Transport Phenomena** has its own transversal validity along almost any engineering specialization. Along with their physical, analytical, and numerical frameworks (even at the present introductory level), students become more adaptable and versatile to succeed in a knowledge-based global marketplace. This book may represent a starting thrust to design, simulation, and optimization of many kinds of industrial and environmental process modeling.

**Framework** The sequence of the arguments is fairly close to what is traditionally covered in beginning Transport Phenomena courses. A course on engineering thermodynamics is a prerequisite.

Each transported quantity is described based on its *driving force* and *flux*, as is usually suggested by classical and contemporary references [1, 2]. The media in which these quantities are transferred are considered as continuous, and their molecular bases are discussed. Phenomenology is described based on its analogies, while the symbolic structure of its computational modeling is hinted at, along with some additional ideas on effective modeling. At the end, the reader will learn on what is "behind the scene" in popular software that solve for Transport Phenomena applications, even with combined mechanisms involved.

In Chap. 1, the study of the discipline is introduced, with emphasis on the various modes of heat transfer as a multiphysics framework. Chaps. 2–5 cover the bases of transport of quantities that are *conserved in nature*, from heat to chemical biological species, through momentum.[4] Finally, in Chap. 6, a number of contemporary applicative cases are presented.

**Useful features** In lieu of completeness, I chose synthesis and presentation clarity. To facilitate learning, previously introduced equations are not simply recalled by their number but they are fully reported in the text instead, along with their equation number, so that the reader will instantly recognize their notation and avoid chase them in the preceding pages.

Basic calculus notations (with special reference to differential equations) have been recalled whenever necessary.

Care has been exercised in using ultra-clear graphical contents and in choosing the most appropriate symbol sets. In particular, all driving variables, properties, and parameters (whenever their variation is implied in the context) are in italics, whereas all other symbols are upright. This rule reflects in the choice of subscripts and superscripts, as well. As an example:

---

[3]Later on in the book, these frameworks are referred to as combined Physical/Chemical/Biological processes (PCB).

[4]The transport of a fourth quantity, the electric charge, has not been covered, along with the entire topic of radiation heat transfer.

- a variable such as the temperature of a process is denoted by $T_\mathrm{p}$, while the constant-pressure specific heat will be $c_p$
- on the other hand, $T_P$ is the value of temperature at the point $P$, and $T_E$ is the corresponding value when we consider the neighbour point at the east side of $P$
- the thermal fluxes which differ for their physical mechanism, whether conductive or convective, are referred to the coefficient of distinguishing law for the specific mechanism: so they will be $\dot{q}_\lambda$ or $\dot{q}_h$, respectively
- convective heat or mass transfer coefficients will be labeled corresponding to the mechanism they refer: so they will be $h_\mathrm{T}$ or $h_\mathrm{M}$, respectively; while v or t will denote laminar (viscous) or turbulent quantities in subscript or superscript mode
- dimensionless quantities are always denoted with upright characters

**Limitations** As implied above, in this dense but compact reference, an introduction of the above subjects is accounted for, only. Deeper development of Transport Phenomena can be found in the books of broader breath and completeness, as indicated in the bibliography. Those works have been taken as landmarks for the development of the present material and represent a source of continued inspiration throughout my activity.

I would like to thank in advance the ones who will report any inaccuracies or errors, thus helping to make this book a valuable learning aid.

Potenza, Italy                                                                             Gianpaolo Ruocco
April 2017

# Acknowledgements

I wish to thank the following Colleagues for their support and suggestions during the preparation of the manuscript:

P. Caccavale, University of Naples "Federico II"
M.V. De Bonis, University of Basilicata
F. Marra, University of Salerno
A. Mulet Pons, Universitad Politécnica de Valencia
F. Murena, University of Naples "Federico II"
S. Nardini, University of Naples "Luigi Vanvitelli"

The encouragement of Prof. Ida Bochicchio and Prof. Ermenegildo Caccese, both at the University of Basilicata, is gratefully acknowledged. My gratitude extends to Mr. Beppe Ruocco, for his editing assistance and insights.

# Contents

# Acronyms

BC      Boundary Condition

CFD      Computational Fluid Dynamics: the framework of application of fluid mechanics to engineering problems

*CS*      Control Surface: an imaginary surface enclosing the thermodynamic system at stake. It can superimpose with real surfaces. It is indicated with a dashed line and envelops the Control Volume

*CV*      Control Volume: The volume in which the mass of the thermodynamic system at stake is contained. It is delimited by a Control Surface

FE      Finite Element method: a discretization method based on the *calculus of variations*, for which solving a PDE is equivalent to minimizing a related quantity called the *functional*

FV      Finite Volume method: a discretization method with which the calculation domain is divided into a number of non-overlapping smaller finite volumes

HEX      Heat exchanger: a device hosting fluid streams whose temperature changes as a result of heat transfer from/to the device walls

NTU      Number of Transfer Units: a dimensionless parameter that qualifies the heat transfer capability of a heat exchanger

OoM      Order of Magnitude

PCB      Physical/Chemical/Biological: the three frameworks in which a multidisciplinary analysis can be brought over. More specifically, PCB is associated with *physical* effects such as phase-changes, *chemical* effects such as reactions, *biological* effects such as metabolism and evolution

PDE      Partial Differential Equation

RANS      Reynolds-averaged (time-smoothed) Navier–Stokes equations

SI          International System of Units (in French: Système International
            d'Unités): It is the modern form of the metric system and is the world's
            most widely used system of measurement
SIMPLE      Semi-IMplicit Pressure Linked Equations method: a basic
            pressure-based method, widely used by the incompressible flow
            modeling community

# Chapter 1
# Transport Phenomena and Multiphysics Modeling

**Abstract** The motivation and principles of the book are outlined. The role of transport phenomena is presented, in the framework of actual and virtual experiments, with the focus on process engineering. The heat transfer, as an example of multiphysics transport, is briefly introduced with its different modes.

## 1.1 Motivation of This Book

This book finds its motivation as **transport** or **transfer phenomena** which can be observed in a great variety of practical situations of the process industry, as well as in the natural environment, pervading many aspects of our life. Phenomena such as heat and mass transfer, fluid flow and chemical/biological reactions/transformations are often found to bridge among Physics, Chemistry, and Biology (PCB, or *multiphysics* in short). In addition to their strong inter- or multidiscipline character, the phenomena at stake here share many formal aspects and behavior, so that their implementation in different frameworks is possible and desirable, starting from with the classical approach [1].

In many situations or applications, the deepening of knowledge and analysis of combined issues, such as these, is desirable, but frequently it involves considerable hurdles. An alternate methodology based on *mathematical modeling* allowing for systematic prediction is therefore necessary. When based on experimental investigation, these predictions often demand considerable costs, while the alternative method applied to relevant governing variables, described in the continua, offers a number of advantages.

In all processes in which a given entity is being transported or transferred, one can speak of transport phenomena; the higher is the product value for a given commodity sector or cost for an environmental situation, more important and meaningful is the study of transport phenomena to analyze, verify, and optimize the various aspects of the medium being conditioned by such phenomena.

© Springer International Publishing AG 2018                                                         1
G. Ruocco, *Introduction to Transport Phenomena Modeling*,
https://doi.org/10.1007/978-3-319-66822-2_1

## 1.2 Why Study Transport Phenomena?

### 1.2.1 A First Statement on Transport Phenomena

Three main quantities are at stake in this framework:

1. **Heat,** studied by the *heat transfer* branch of physics,[1] is the quantity whose variation in a system that moves or is at rest determines the **distribution of temperature**. Heating and cooling are common operations in substrate processing. Heating on raw media is performed for various purposes, but heat is also transferred often when momentum transfer is present. Heating substrates allow for reduction of the microbial population, inactivation of enzymes, reduction of the amount of water, and modification of the functionality of certain compounds. On the other hand, heat is removed from substrates to reduce the rate of its deteriorative chemical and enzymatic reactions and to inhibit microbial growth, extending commercial life by cooling and freezing. Knowledge is needed to achieve better control and avoid under- or over-processing, which often results in detrimental effects on media characteristics.

2. **Momentum,** studied by the *fluid dynamics* branch of physics,[2] is the quantity whose variation in a fluid in motion determines the **distribution of velocity** in the fluid itself. Momentum transfer is a common mechanism in many processes; very often, heat and/or mass transfer phenomena occur in association with flow, such as in heat exchanger flows. Often, with momentum transfer at stake, we are driven to study different *phases* at once, so the effect of the phenomena realizing on their *interface* is of special interest. As an example, when analyzing the fluid field with concomitant phenomena, it is advised to adopt the continuity of energy and mass across the solid/fluid interface (conjugate modeling). Some particular considerations should be taken into account when dealing with momentum transfer in the applied sciences: chemical components and some rheological property play important roles in each manufactured product.

3. **Mass** (in the sense of *chemical or biological species*, or *species* in short), studied by the *mass transfer*, is the quantity whose variation in a system determines the **distribution of species concentration**. Mass transfer is the migration of a substance through a mixture under the influence of a concentration gradient in order to reach equilibrium: this applies in a broad sense to chemical or biological species. A substance can also be convected by a flow field (when momentum transfer is concomitant), which happens in a number of processes. Biochemical and chemical engineering operations, and all separation techniques involve mass transfer. In substrate processing, mass transfer phenomena are present, for example, in drying and transmission of vapor and gas through a film or mem-

---

[1] Heat transfer is the exchange of thermal energy between physical systems, depending on temperature and pressure.

[2] Fluid dynamics is a subdiscipline of fluid mechanics that deals with fluid flow: the natural science of fluids (liquids and gases) in motion.

brane. Substrate stability and preservation are also affected by mass transfer of environmental components that can affect the rate of related reactions.

In this book, each of these quantities will have a dedicated chapter, and the link between them will be addressed in the *medium/process/system* combination,[3] by a later chapter dealing with their application in a variety of unit operations.

## *1.2.2  Transport Phenomena in the World Around Us*

In process engineering, a *unit operation* consists in a single PCB transformation to manufacture or condition a given medium or product, by means of a given process being hosted in a system or plant. Unit operations can deal with more transport phenomena at once and consist of five classes:

1. Fluid flow processes, including fluid transportation, compression/expansion/pumping, extrusion, filtration, fluidization, mixing, and atomization.
2. Heat transfer processes, including evaporation and heat exchange.
3. Mass transfer processes, meaning separation between species or phases, including adsorption,[4] absorption,[5] desorption,[6] concentration/condensation, distillation, drying/evaporation/humidification, extraction, precipitation, and osmosis.
4. Thermodynamic processes, including gas liquefaction and refrigeration.
5. Mechanical processes, including solids transportation, crushing and pulverization, and screening and sieving.

This book concerns the first three classes, where some unit operations combine among these groups. The study of transfer phenomena carries several outcomes:

- knowledge of the process
- design and/or verification of experimental techniques
- deduction of empirical correlations
- set-up of process simulations

Transport phenomena can be revived in a number of frameworks, among others:

- agriculture, environmental engineering, meteorology
- biology, biotechnology
- chemical engineering, pharmacy
- food technology

---

[3]This framework can be regarded as a *scale triad*, as the physical range of phenomena existence spans when passing from one nesting component to the other. A very important procedure in engineering, the scaling up, is related to this triad.

[4]Adhesion of atoms, ions, or molecules from a gas, liquid, or dissolved solid to a surface.

[5]Inclusion of particles of gas or liquid in liquid or solid material.

[6]Release of a substance from or through a surface. It is the opposite of adsorption/absorption.

- material science
- mechanical engineering
- physiology

In the last chapter of this book, a personal account of some applications drawn from these frameworks will be presented, in form of case studies.

## 1.3   A Vision of Transport Phenomena Analysis

### 1.3.1   Experiments Versus Virtualization

As already implied above, prediction of heat and mass transfer and fluid flow can be usually obtained by two methods: experimental and computational investigations.

The most reliable source of information about a PCB process is often represented by actual measurements. Experiments can involve full-scale equipment and systems and can be employed to predict the course of the process in different conditions. However, full-scale tests are usually expensive or often impossible, for the entire range of variables involved. The alternative is to resort to small-scale rigs or systems, whose results must then be scaled-up. On the other hand, the operation on small-scale systems sometimes misses important features, which reduces the benefit of the information gathered. Finally, we must keep in mind that experimental procedures are not free from errors.

On the other side, analytical investigations allow one to work out the consequences of a mathematical model. In the present framework, this generally consists in a set of partial differential equations (PDEs) stemming from conservation principles to represent the actual PCB process to its full extent, but analytical (exact) predictions require solutions in closed form, which are way too complex and often prohibitive. Therefore, we need to turn to approximated formulations represented by computational or numerical modeling, in the end consisting in solving a large *system of algebraic equations*.

So, as more sophisticated software tools (complemented by adequate hardware) become available, more complex problems can be approached by numerical simulation: modeling and simulation are successfully present at many levels in education, research, and production. "Virtual experiments" bring detailed and complete information, complementing nicely with PCB experiments: all relevant features can be probed, with no disturbance by the measuring tool, even in those locations that are normally inaccessible. Moreover, the cost of a computer run is a tiny fraction of the corresponding PCB experiment and can also be accomplished with remarkable speed, so that many different configurations of the same process can be usually tried in a much shorter period.

The broad field of numerical modeling covers the range from the automation of well-established engineering design methods, to the use of detailed solutions of the driving equations as substitutes for experimental research, into the nature of complex situations. One can purchase design packages to solve simple problems in a few seconds on a personal computer or may deal with an integrated task involving hundreds of computing hours narrowing down to the slightest product/process/system triad detail.

## 1.3.2 Learning Transport Phenomena Through a Recursive Approach

Unfortunately, systematic application of transport phenomena in the current technological practice is frequently missing, as empiricism or "pragmatical approach" still dominates the scene.[7] Transport phenomena are poorly understood by non-specialized personnel, due to the inherent mathematical difficulties and the lack of coverage that is given the subject during their professional training. Moreover, the classic theory of this subject is applied with some difficulty to biological materials, due to their peculiar character (structure, properties variation depending on governing variables, etc.).

In this book, an elementary treatise, based on the analogy theory of transport phenomena [1], is proposed by using a multidisciplinary approach which materializes when the same formal and solution tools are employed, facilitating the management of modeling and computational segments. To this end, the same topic sequence has been adopted for each of the chapter treating each of the main transport quantities. Among these, heat deserves some special remarks, as it can be transported more easily regardless of the phase. In other words, in Chap. 2, the transport of heat in solid or fluid at rest is examined, but the same quantity can be transported in fluids by means of momentum, which is explained in Chap. 3. As a consequence, in Chap. 4, the transport of heat is studied again, but with the momentum transport underlying, this time. The reason of this distinction is that due attention is given to the transport of heat and mass when phases interact, as in a fluid flow being deformed by the presence of a solid object: it is clear then that Chaps. 3–5 do deal with such an occurrence, which does not take place for pure heat transport, in Chap. 2.

Figure 1.1 introduces to the *sequence of topics which has been recursively adopted* for each phenomenon. We begin with each molecular mechanism, featuring material properties and coefficients that specifically play a role. Then, few selected elementary configurations are surveyed, while a conservation balance is called upon to introduce to the fundamental or governing PDEs. These can be purposely rendered dimensionless, by opportunely grouping some paramount variables with dimensional analysis. Basic interactions between different materials or phases are examined to focus on the importance of combined mechanisms in multiphysics, such as the flow defor-

---

[7]One common example is the use of empirical notations as average transfer coefficients, e.g., applied at the external surface of a substrate being exposed to a working fluid. This is a limitation which needs addressing through *conjugate* modeling as implied earlier, regardless of the phases interface, solving the energy/mass transport in both phases simultaneously.

**Fig. 1.1** Sequence of the topics covered by each Chaps. from 2 to 5

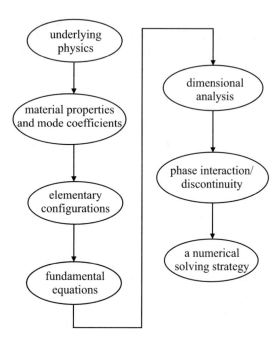

mation associated with thermal/composition perturbations in a given fluid streaming past a solid surface. Finally, the fundamentals of the numerical solution strategy is explained in each chapter, while few application examples are carried out in Chap. 6, to illustrate and discuss some contemporary modeling practice.

Based on transport phenomena knowledge, process virtualization has its importance in key-enabling technologies such as the advanced material science, manufacturing, and processing. Many modeling difficulties can be attacked such as the *coupling* among different transport mechanisms.[8] As an instrument in the multidisciplinary approach to production processes, the exercise of virtualized transport phenomena will find its role in many research and innovation action, as inspired, for example, by the Industrial Leadership Section of the European Commission's Horizon 2020 program[9] and, besides, will represent a challenging area for future research and development for decades ahead worldwide.

---

[8]This arises as transport phenomena can be easily be intertwined, i.e., interdependent—e.g., when liquid water evaporates from a heated substrate, producing vapor water at the expenses of the energy budget.

[9]As deduced by http://ec.europa.eu/programmes/horizon2020/en/h2020-sectionleadership-enabling-and-industrial-technologies. Cited 15 Oct 2015.

## *1.3.3 A General Governing Differential Equation*

The common engineering way to express the governing laws of transport phenomena is by means of PDEs, representing a certain conservation principle: if a general quantity $\phi$ is cast as its dependent variable, the equation implies that there is a balance among the various terms. $\phi$ will depend on time and one or more geometry coordinates (from *1-D* to *4-D situations*); moreover, the process may contain or not the dependence on time, the situation being called an unsteady or steady state, respectively.

The dependent variables are usually *specific properties*, i.e., quantities expressed on a *unit-mass basis*,[10] but the equation terms themselves are given on a *unit-volume basis*. One can ascertain this fact by carrying out a dimensional analysis on the terms forming the general transport PDE for variable $\phi$:

$$\frac{\partial \rho_\phi \phi}{\partial \theta} + \nabla \cdot \left( \rho_\phi \mathbf{w} \phi \right) = \nabla \cdot \left( D_\phi \nabla \phi \right) \pm S_\phi \qquad (1.1)$$

where $\theta$ is *time* and $\mathbf{w}$ is the *velocity vector*, and $\rho_\phi$ and $D_\phi$ are the *density* and *diffusion coefficient* for the specific $\phi$. For example, setting $\phi$ equal to specific enthalpy $h$, and knowing that the *nabla operator* $\nabla$ is dimensionally equal to $(1/m)$[11] in the International System of Units (SI), one gets a mass density $\rho$ (kg/m$^3$) and a heat diffusion coefficient $D = \lambda/c$, with thermal conductivity $\lambda$ (W/mK) and specific heat $c$ (J/kgK), as will speculated later in Chap. 2.

Special attention must be devoted to *source/sink term* $S_\phi$ that will materialize in various forms depending on the nature of $\phi$, which is *paramount in a multiphysics framework*, specially with intertwined PCB occurrences, as with processes to biomedia.[12] It is purposely Eq. (1.1) that we intend to **transform in a suitable way to carry out product/process virtualizations** in the situations at stake.

## 1.4 Heat Transfer as a Multiphysics Framework

The analysis of how heat flows brings over an excellent opportunity to consider that various physics mechanisms contribute to a number of common phenomena. Whenever a *temperature gradient* in a system exists, or two systems at different temperature are in contact, one has **transfer or flows of energy in the form of heat**. This notion has been nurtured since the study of Thermodynamics, that foregoes the topics at stake here. The flow of heat cannot be directly measured or observed, but

---

[10]Such as specific enthalpy, a velocity component, or a concentration of a chemical species.

[11]A common notation to report on a entity dimensions exploits square brackets, i.e., $[\nabla] = 1/m$.

[12]See, for example, De Bonis, M.V., Ruocco, G.: Computational Transport Phenomena in Bioprocessing with the Approach of the Optimized Source Term in the Governing Equations. Heat and Mass Transfer (2012). https://doi.org/10.1007/s00231-012-0992-z.

**Fig. 1.2** A control volume,
$CV$, and its confining control
surface, $CS$

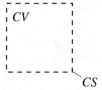

the effects produced indeed can. With this notion, the flow of heat (as for the other form of energy, the mechanical work) through a control surface ($CS$) of a system implies a variation of its energy content in a control volume ($CV$, Fig. 1.2). All such processes which involve energy exchange and conversion must, therefore, obey to the First and Second Law of Thermodynamics, but the specific laws of heat transfer cannot be deduced from these laws: the approach generally taken by the engineering Thermodynamics allows one to study only the equilibrium states. Consequently, as the flow of heat is due to an imbalance of temperature, its quantitative analysis must ground on a different discipline.

### 1.4.1   The Role of Heat Transfer

From an engineering point of view, once a $CS$ and a $CV$ are assigned, the essential problems of heat transfer are the knowledge of *thermal power transmitted* through the $CS$ and the *temperature distribution* on the $CS$ or in the $CV$ [2]. Many quantities and parameters do indeed depend on temperature and exchanged heat, in the treatment of a number of *fluids and solid media*. The detailed analysis of heat transfer in the process to assess the cost, the application range, and size of the hosting plant is frequently paramount to applicative cases. Time enters in the analysis as we may want to *exchange heat in a given time interval*. Boilers, radiators, reactors, and heat exchanger sizes depend not only on the *quantity* of exchanged heat, but primarily on the *rate* at which the heat is exchanged in the assigned conditions. Moreover, the proper working of system elements depends on heating and cooling means that are attributed, based on the design, by *continually transferring at high rate heat* through the system surfaces. Therefore, in heat transfer problems one has to identify whether the process operates in a **steady** or an **unsteady** (transitory) **state**:

- When the internal energy in the $CV$ does not change in time, temperature is uniform and steady conditions are established: in any point in the $CV$, the heat power entering equals the heat power exiting, and no variation of energy occurs.
- Conversely, if temperature does change with time, unsteady conditions occur. The temperature variation suggests that the internal energy content is changing in the system, depending on time-dependent heat fluxes across system's $CS$.

**Fig. 1.3** An informative scene on different fire extinguisher practices, i.e., on the different heat transfer modes (① to ③)

## 1.4.2   The Different Modes of Heat Transfer

Heat exchange is not ruled by a single relationship but rather by a combination of independent physical laws or mechanisms, describing three different heat transfer modes:

① **conduction** (or *diffusion/microscopic transport*)
② **convection** (or *macroscopic transport*)
③ **radiation**

Let us consider the scene depicted in Fig. 1.3, in which a group of people (representing *the considered medium* or *substrate*) is working in a coordinated effort by using tap water (*the heat being transmitted through the substrate*, in the arrows direction) to extinguish a raging bonfire (the difference between the tap water coming from the wall tap and in the one contained in the fire represents the *temperature difference*[13]).

①  We see that this group of people passes hand-in-hand a water bucket: this is equivalent to the **heat transfer by conduction**. With this, heat flows from hotter

---

[13]This quantity is also called the potential difference, or the driving force for the heat flux.

to colder regions through a single medium or through different media that are in *direct contact*. The energy is transmitted this way among the molecules, with no need for molecule motion. According to the kinetic theory, temperature is proportional to the average kinetic energy of a medium's molecules, and the internal energy represents the sum of all microscopical energy contributions associated with such molecules. When some molecules in a region of the medium reach a kinetic energy level higher than the one of an adjacent region, as indicated by a temperature difference, the molecules having more energy give way to some of it to the region at lower temperature.[14] The detectable effect of conduction is that the temperature levels out. Yet, if temperature differences are maintained due to supplied or dissipated heat, a uniform heat flux is established from hotter to colder regions.

② Let us consider then that some men can run, each carrying a bucket, from the supplying tap to the fire: this is equivalent to **convection heat transfer**. With this, heat flows from hotter to colder regions by means of a combined action of *conduction, variation of energy content, and mixing*: the most important heat transfer mode between a solid surface and a fluid. For example, from a hotter surface heat is first transferred by conduction to adjacent fluid molecules (causing an increase in internal energy and temperature in the fluid) and then the heated molecules move out to colder fluid regions and mix with them, releasing some of their energy to other molecules. Thus, we have a *combined flux of matter and heat*; that is, the energy is collected by the moving molecules and dissipated with them by their motion. We can distinguish between **free** or **natural convection** and **forced convection**, according to the cause that drives the motion. In the former, motion depends solely on *density differences due to temperature gradients*; in the latter, motion is induced by some *external factor*, such as pump or fan. Heat exchange processes involving *phase-change* can be taken as convective due to the induced motion: as in the bubbles ascent during evaporation of drops fall during condensation. As the efficiency of the heat transfer greatly depends on fluid motion, the study of convection is based on the knowledge of its fluid dynamics features .

③ Finally, the water can be launched over the fire by means of a proper pump-driven nozzle, feeding directly from the wall tap, *independently on the medium passed through* (no people are involved in the water launch): this is equivalent to the **radiation heat transfer**. With this, heat flows from hotter to colder regions *with no contact*, even with interspersed vacuum. The term "radiation" generally refers to any propagation of electromagnetic waves, but at stake here are temperature-dependent phenomena, allowing the energy transport through transparent media or vacuum. All bodies emit continuously heat by radiation, with intensity that depends on temperature and surface state. The radiant energy travels at light speed ($3 \times 10^8$ m/s) and presents the same features than *light*: this is nothing more than radiation which is visible to the human eye, having frequencies that

---

[14]In fluids, these happen due to elastic impacts; in metal solids, this is due to the diffusion of electrons.

belong to a subset spectrum of the wider heat radiation. This is why radiation heat transfer can be studied based on the wave theory. According to this, the radiated heat is emitted from a body by finite quantities or energy *quanta*. With analogy with light radiation, heat is subject to both interference (wave nature) and photoelectric effect (particle nature). When radiation hits another body, its energy gets absorbed only in the vicinity of the exposed surface. Heat transfer becomes more important with body temperature increase, but for those applications in which the temperature is not too far from the ambient, radiation heat transfer can be neglected altogether.

### 1.4.3  Multiphysics in Heat Transfer

Only conduction and radiation can be classified as heat exchange processes, as these specific physical mechanisms directly depend on a temperature difference. On the other hand, heat convection is a combination of momentum transfer and conduction, but its analysis has been associated with the other two modes as it results in mean or *bulk*[15] heat flow from hotter to colder regions, with the same motivation. Each of the three modes can be described and studied separately, but in many cases, **heat flows by more than one mode, simultaneously**. One must then assess the relative importance of these modes, as in practice, when one mode is predominant over the others, it is useful to neglect these at least until the operating conditions change so that the inclusion of more modes is suggested.

In this book, we will not delve into radiation heat transfer, as the phenomenological laws are completely different from the other mechanisms and its analysis cannot be treated by using the PDEs coming from conservation principles. In-depth treatment of this topic, along with its interactions with conduction and convection, can be found in specialized works [3].

## 1.5  Further Reading

- The nature of transport phenomena. Thompson, W.J.: Introduction of Transport Phenomena. Chapter 1. Prentice Hall, Upper Saddle River (2000)
- Available applications of heat and mass transfer in the biological and environmental sciences. Datta, A.K.: Heat and Mass Transfer—A Biological Context. CRC Press, Boca Raton (2017)
- Scaleup in restrospect. Astarita, G.: Scaleup: Overview, Closing Remarks, and Cautions. In: Bisio, A., Kabel, R.L. (Eds.) Scaleup of Chemical Processes. Wiley, New York (1985)

---

[15]The term "bulk" is employed to mean a spatial average variable (usually an area-weighted one): in this case, across the flow cross section.

- Radiation. Modest, M.F.: Radiative Heat Transfer. Academic press, New York (1993)
- Radiation. Siegel, R., Howell, J.R: Thermal Radiation Heat Transfer. Hemisphere, New York (1992)


## References

1. Bird, R.B., Stewart, W.E., Lightfoot, E.N.: Transport Phenomena. Wiley, New York (2002)
2. Bergman, T.L., Incropera, F.P., Lavine, A.: Fundamentals of Heat and Mass Transfer. Wiley, New York (2011)
3. Özışık, M.N.: Radiative Transfer and Interactions with Conduction and Convection. Wiley, New York (1973)

# Chapter 2
# Heat Transfer by Conduction

**Abstract** Being this the first mode that traditionally is encountered in the study of heat transfer, with the analysis of *conduction* the ground will be set for more complicated transfer phenomena. After a brief reference to the basic physical mechanism, as we recognize that **heat transport in stationary media** is driven by a *temperature difference*, we start by exploiting first the *macroscopic balance* for heat conduction. Then, the opportunity is seized to develop proper analytical skills by deriving and integrating the *governing differential equations* in various cases, following the *microscopic balance* leading to the **distribution of the temperature scalar**. With this mechanism, the subject medium participates only through the heat source or sink. Next, some graphical tools will be presented that are of some use for the solution of transient cases, and finally a *numerical solution* of the governing equations is proposed and initiated, to cast the base for a discussion on more complex transfer phenomena.

## 2.1 Conduction: The Underlying Physics and Basic Definitions

### 2.1.1 Molecular Heat Flow

Let us consider a metallic pot holding some liquid on a gas stove at a certain cooking stage. In order to describe the onset of heat conduction and related temperature distribution in the pot, a *thermogram* can be obtained by means of *thermal imaging*, as in Fig. 2.1. The combustion gas ① (white) makes direct contact with the metal, so that heat flows through pot's wall giving rise to a quantitative temperature difference ② (green–hot, dark blue–relatively cooler).

Conduction deals with the three effects applying to heat that is managed when two stationary bodies or media in thermal disequilibrium (even microscopic in size)

---

The original version of this chapter was revised: Belated corrections have been incorporated. The correction to this chapter is available at https://doi.org/10.1007/978-3-319-66822-2_7

© Springer International Publishing AG 2018

G. Ruocco, *Introduction to Transport Phenomena Modeling*, https://doi.org/10.1007/978-3-319-66822-2_2

**Fig. 2.1**  A thermal imaging
measurement of hot gas from
a kitchen stove ① and of
pot's wall and contained
liquid ②

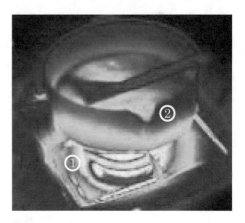

are in intimate contact: the *net transfer of heat*, the *thermal inertia*, and the *internal heat generation/dissipation*. The first two effects are governed, beside the driving force for the heat flow, by the medium thermal properties, and the third effect being is governed by the inherent *multiphysics* instead.

Let us start our analysis by taking a closer (microscopic) look at the medium.

## 2.1.2  Fourier's Law of Conduction

### 2.1.2.1  Temperature Profile in a Conductive Medium

Let us consider a $CV$ consisting in a layer of a conductive medium or substrate (solid, or fluid at rest) enclosed by two parallel plates, as in Fig. 2.2a. The substrate is defined by a finite thickness $L$ along $y$, while it is very wide along the other two coordinates. This large width (area), in any $x - z$ plane, will have a size $A$. The substrate and the plates are initially in thermal equilibrium at $T_0$.[1] At a time $\theta = 0$, the lower plate is set and maintained at a higher temperature $T_p$ (Fig. 2.2b).

A temperature profile or distribution $T(y, \theta)$, variable in space and time, is therefore created as in Fig. 2.2c due to **molecular heat transport**: after some time (Fig. 2.2d), when the steady state is restored, the substrate's molecules that touch the bottom plate are found at the same $T_p$, while those in contact with the top plate still stay at the same initial $T_0$.[2] In other words,

$$T(0, \infty) = T_p, \quad T(L, \infty) = T_0 \tag{2.1}$$

---

[1]The thick line in Fig. 2.2a is normal to the $T$ axis.

[2]The temperature profile will be linear with $y$ or otherwise, depending on the material characteristic with respect to conduction.

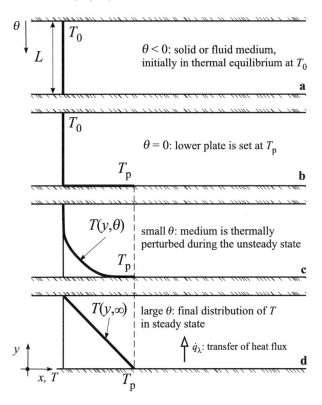

**Fig. 2.2** Variation, during time progress $\theta$ (from top **a**, to bottom **d**), of the temperature profile $T$ in the medium, due to the thermal perturbation given by the hotter lower plate. With reference to Note 2, in this case we observe a linear profile. The direction of the thermal flux $\dot{q}_\lambda$ is evidenced at lower right

### 2.1.2.2 Flow of Heat by Conduction

Let us denote with $\dot{Q}_\lambda / A$ the **conductive thermal flux**, that is, the thermal power $\dot{Q}_\lambda$ transferred by conduction along the $y$-direction through the layer surface $A$, applied to the lower plate in order to maintain the temperature difference or gradient $\Delta T = T_p - T_0$ across the substrate thickness. For the situation reproduced in Fig. 2.2d, it is therefore:

$$\frac{\dot{Q}_\lambda}{A} \equiv \dot{q}_\lambda \propto \frac{\Delta T}{L} \tag{2.2}$$

The thermal flux $\dot{Q}_\lambda / A$ causes a constant **net transport of heat through the adjacent molecules**, *in the direction taken by temperature decrease*.[3] Equation (2.2) evidences the existing link between the *heat transferred through the substrate* and

---

[3] Note that this fact agrees with the *Second Law of Thermodynamics*.

the *observable temperature field*. In other words, the **temperature gradient is the driving force of heat transfer**. For the thermal power and the thermal flux, it is $[\dot{Q}_\lambda] = W$ and $[\dot{q}_\lambda] = W/m^2$, respectively;[4] while it is $[L] = m$ and $[\Delta T] = K^5$ or °C, for the thickness and the difference of temperature, respectively.

### 2.1.2.3 Link Between Temperature Difference and Conduction Heat Transfer

We stated already that, once a $CS$ and a $CV$ are assigned, the essential problems in studying conduction heat transfer are

- the *determination of the thermal power transmitted under a given temperature difference* through the $CS$, and
- the *determination of the temperature distribution* on the $CS$ or in the $CV$

The distinctive proportionality parameter, or *material property* inherent to Eq. (2.2), is characteristic of the substrate, and it is called its **thermal conductivity** $\lambda$:

$$\boxed{\frac{\dot{Q}_\lambda}{A} = \lambda \frac{\Delta T}{L}} \tag{2.3}$$

This is the **Fourier's Law of conduction or thermal diffusion**:[6] *the thermal power transferred by conduction through a layer of a substrate is equal to the product of its thermal conductivity, the area of the section through which the heat flows (measured perpendicularly to the direction of the flux), and the temperature gradient through the substrate.* This relationship describing the conduction phenomenon is firstly valid for a solid layer, but also for a liquid or gas layer, provided that other heat transfer modes can be neglected.

In the present framework, a *differential form* (the relationship obtained as the layer thickness $L$ tends to 0) is preferred to the *algebraic form*[7] of Eq. (2.3), except when a *macroscopic balance* is sought (as will be seen later). Having noted that the thermal flux is positive when the temperature gradient is negative, we can introduce the **Fourier's Law of conduction in the steady state in differential form**:

$$\boxed{\dot{Q}_\lambda = -\lambda A \frac{dT}{dy}} \tag{2.4}$$

---

[4]The unit of power, watt, is named after Scottish engineer J. WATT (beginning of the nineteenth century).

[5]The Kelvin scale is named after the Northern Irish mathematical physicist and engineer W. THOMSON, Ist Baron Kelvin (beginning of the nineteenth century).

[6]As proposed by French mathematician and physicist J.- B. FOURIER, in the early nineteenth century.

[7]Also called a discrete form.

The thermal conductivity $\lambda$, as any other material property, may change in any material point in the $CV$. In this case, Eqs. (2.3, 2.4) refer to a situation with *uniform thermal conductivity*.

### 2.1.3 Driving Material Properties

Beside thermal conductivity $\lambda$, the property parameters that will be used in the development of the governing equations for this transfer mechanism are the *specific heat* $c_p$, and the *density* $\rho$, that may change depending upon the thermodynamic state (interatomic distance of matter or aggregation, and temperature and pressure) [1, 2]. Furthermore, these property parameters can vary in other three ways (Fig. 2.3):

1. each property can be *variable with time*, in unsteady processes
2. several values of a given property can also be present at once within a certain portion of a body: this case is called the material *non-homogeneity*
3. each property can also depend on the orientation with respect to the driving force of the phenomenon at stake (in this case, the temperature difference): this case is called the material *anisotropy*.

A convenient way to express a property variation is in terms of a *tensor*.[8]

Frequently, in the analytical development of the governing equations, these three variation effects can be disregarded: for way 1, the property will be taken as *constant*, and for ways 2 and 3, it will be taken as *uniform*. When needed, their variations can be dealt with, in the numerical development of the governing equation, by using a *piecewise-constant* formulation or by iterative techniques, as proposed later in Sect. 2.5 (p. 55).

Common materials feature different properties values. Let us start with $\lambda$, whose order of magnitude (OoM) decidedly increases with reducing interatomic distance (Table 2.1); $[\lambda] = \text{W/mK}$. For many fluids of technical interest, $\lambda$ increases considerably with $T$, while for many solids this variation of $\lambda$ can be neglected[9] (Table 2.2).

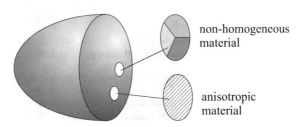

**Fig. 2.3** Beside thermodynamic state and unsteadiness, properties can change in a given body due to non-homogeneity and anisotropy

non-homogeneous material

anisotropic material

---

[8]Tensors are important in physics because they provide a concise mathematical framework for formulating and solving physics problems in areas such as elasticity and fluid mechanics, among others.

[9]For many media in the biosciences, thermal conductivity $\lambda$ will take an intermediate value between those of liquid water and air.

**Table 2.1**  Order of Magnitude (OoM) ranges of thermal conductivity $\lambda$ (W/mK)

| Aggregation state | OoM |
|---|---|
| Gases at standard pressure | $10^{-2}$–$10^{-1}$ |
| Liquids | $10^{-1}$–$10$ |
| Non-metallic solids | $1$–$10$ |
| Metallic solids | $10$–$10^2$ |

**Table 2.2**  Indicative values of thermal conductivity $\lambda$ (W/mK) for water, air, and some solids as a function of temperature $T$ (°C)

| $T$ (°C) | Water, sat.liq. | Water, super.steam | Air | Steel, stainless | Brick, generic | Glass | Ice |
|---|---|---|---|---|---|---|---|
| 0 | – | – | – | – | – | – | 2.22 |
| 20 | 0.597 | – | $2.55 \times 10^{-2}$ | 14.4 | 0.45 | 0.81 | – |
| 100 | 0.682 | $2.40 \times 10^{-2}$ | $3.13 \times 10^{-2}$ | – | – | – | – |
| 150 | 0.670 | $2.80 \times 10^{-2}$ | $3.45 \times 10^{-2}$ | – | – | – | – |

**Table 2.3**  Indicative values of specific heat $c_p$ or $c$ (J/kgK) for water, air, and some solids as a function of temperature $T$ (°C)

| $T$ (°C) | Water, sat.liq. | Water, super.steam | Air | Steel, stainless | Brick, generic | Glass | Ice |
|---|---|---|---|---|---|---|---|
| 0 | – | – | – | – | – | – | 1830 |
| 20 | 4182 | – | 1006 | 461 | 840 | 800 | – |
| 100 | 4211 | 2060 | 1011 | – | – | – | – |
| 150 | 4356 | 1963 | 1025 | – | – | – | – |

Values of $c_p$[10] do not show large variation with $T$, while varying decidedly with the solid type[11] (Table 2.3); $[c_p] = $ J/kgK.

Values of $\rho$ for many fluids show an inverted dependence on $T$ (Table 2.4) with respect that for $c_p$; $[\rho] = $ kg/m³. Together with thermal conductivity and specific

---

[10] Strictly speaking, pressure-constant and volume-constant specific heats arise, when dealing with the variations of internal energy $e$ and enthalpy $h$: for example, for an ideal gas it is

$$de \equiv c_v dT \quad dh \equiv c_p dT$$

For solids and liquids, as incompressible matter, there is no distinction between specific heats at constant pressure and constant volume; therefore, the subscripts can be omitted: $c_p \equiv c_v \equiv c$. For an incompressible liquid it is

$$de \equiv c dT \quad dh \equiv c dT + \frac{V}{m} dp$$

with $V$, $m$, and $p$ the volume, mass, and pressure, respectively, at the given thermodynamic state.

[11] For many substrates in the biosciences, specific heat $c_p$ will take an intermediate value between those of liquid water and air.

**Table 2.4** Indicative values for density $\rho$ (kg/m$^3$) for water, air and some solids as a function of temperature $T$ (°C)

| $T$ (°C) | Water, sat.liq. | Water, super.steam | Air | Steel, stainless | Brick, generic | Glass | Ice |
|---|---|---|---|---|---|---|---|
| 0 | – | – | – | – | – | – | 913 |
| 20 | 998.2 | – | 1.20 | 7817 | 1800 | 2800 | – |
| 100 | 958.4 | 0.586 | 0.94 | – | – | – | – |
| 150 | 910.0 | 0.525 | 0.74 | – | – | – | – |

heat, density contributes to form the **thermal diffusivity** $\alpha$ (m$^2$/s) that measures the *propagation of the thermal perturbation*:

$$\alpha \equiv \frac{\lambda}{\rho c_p} \tag{2.5}$$

### 2.1.4  Fourier's Law Generalization

Using a proper subscript to the direction of the thermal flux, a general form of Eq. (2.4) can be written (with uniform $\lambda$), by assuming that $T$ may depend on more coordinates:

$$\dot{q}_{\lambda y} = -\lambda \frac{\partial T}{\partial y} \tag{2.6}$$

Generally, then, the thermal flux is a vector quantity. Let us assume that a general $T$ distribution exists, for example, in rectangular coordinates: $T = T(x, y, z, \theta)$ .[12]

Let us take the surface that within the substrate connects all points having the same temperature: an **isothermal surface**. Then, let us consider two such surfaces, at a given time, having $T$ and $T + dT$ temperatures, respectively, as well as the point $P$ on one of these (Fig. 2.4); the *heat that flows by conduction, in the unit of time, by unit area of isothermal surface* is the **thermal flux** $\dot{q}_{\lambda n}$ **along the general n-direction**. For an *isotropic substrate*, $\dot{q}_{\lambda n}$ will be *normal* to the higher isotherm surface and *oriented toward the lower isothermal surface*, while its *magnitude is proportional to the temperature gradient in the* **n** *direction*.

---

[12]Beside the familiar rectangular coordinate system (also called a *Cartesian* coordinate system, as proposed by French philosopher and mathematician R. DESCARTES in the early seventeenth century), other frequent coordinate systems are the cylindrical system (of length $L$ and radius $R$) and the spherical system (of radius $R$), as reported here at right with the representations of point $P$ and its coordinates.

**Fig. 2.4** Plane projection of
two isotherms, which differs
by an infinitesimal value $dT$,
and of vector $\dot{q}_{\lambda n}$

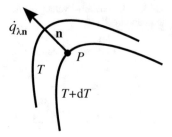

*In scalar terms*:

$$\boxed{\dot{q}_{\lambda n} = -\lambda \frac{\partial T}{\partial \mathbf{n}}}$$

(2.7)

while *in vector form*[13]:

$$\boxed{\dot{\mathbf{q}}_\lambda = -\nabla \lambda T}$$

(2.8)

which is called the **generalized Fourier's Law** *for an isotropic substrate, not necessarily homogeneous.*

In rectangular Cartesian coordinates, the three components of $\dot{\mathbf{q}}_\lambda$ for uniform thermal conductivity are as follows:

$$\dot{q}_{\lambda x} = -\lambda \frac{\partial T}{\partial x} \quad \dot{q}_{\lambda y} = -\lambda \frac{\partial T}{\partial y} \quad \dot{q}_{\lambda z} = -\lambda \frac{\partial T}{\partial z}$$

(2.9)

## 2.2   Elementary Conduction: The Electricity Analogy

In some cases in the practice, when the geometry and the thermal driving forces are simple, the temperature profiles and other related results can be directly computed based upon Fourier's Law of conduction. The simplest of such cases is the **heat**

---

[13]When applied to a scalar, $\nabla$ gives its *gradient* (i.e., a vector). When using a rectangular 3-D coordinate system:

$$\nabla \equiv \frac{\partial}{\partial x}\mathbf{i} + \frac{\partial}{\partial y}\mathbf{j} + \frac{\partial}{\partial z}\mathbf{k}$$

In cylindrical coordinates, it is

$$\nabla \equiv \frac{\partial}{\partial r}\mathbf{r} + \frac{1}{r}\frac{\partial}{\partial \phi}\boldsymbol{\phi} + \frac{\partial}{\partial z}\mathbf{z}$$

In spherical coordinates, it is

$$\nabla \equiv \frac{\partial}{\partial r}\mathbf{r} + \frac{1}{r}\frac{\partial}{\partial \phi}\boldsymbol{\phi} + \frac{1}{r \sin \phi}\frac{\partial}{\partial \theta}\boldsymbol{\theta}$$

The disposition of versors are those illustrated by the axes directions in the schemes of Note 12.

**Fig. 2.5** A wall which is infinite along the y- and z-directions, but with thickness $L$ along $x$: indication of the internal temperature distribution $T(x)$, under a temperature difference $T_1 - T_2$ between the limiting faces

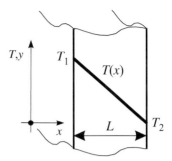

**flow in the steady state through a 1-D (plane) geometry CV**, which will give the opportunity to define some important concept; knowing the temperature distribution is important to evaluate their response to the thermal driving force.

In these situations, the heat flow, or thermal power $\dot{Q}_\lambda$, can be computed by using its definition applied to a plane wall or plate, Fig. 2.5:

$$\dot{Q}_\lambda = -\lambda A \frac{dT}{dy} \tag{2.4}$$

Separating variables one has:

$$\frac{\dot{Q}_\lambda}{A} \int_0^L dx = -\int_{T_1}^{T_2} \lambda dT \quad \text{with} \quad T_1 > T_2 \tag{2.10}$$

The integration limits are determined by observing that the left wall's face at $x = 0$ is at uniform temperature $T_1$, with the right face at $x = L$ being at $T_2$. If $\lambda$ is not dependent on $T$, one has:

$$\dot{Q}_\lambda = \frac{\lambda A}{L}(T_1 - T_2) = \frac{\Delta T}{L/A\lambda} \tag{2.11}$$

In this case, the distribution of temperature in the medium is linear with $x$ (Fig. 2.5).

### 2.2.1 Resistance and Conductance

The temperature difference $\Delta T$ is the *potential* (or driving force) determining the heat flux; the **thermal conductive resistance** $R_\lambda$ that the wall offers to the thermal flux will be

$$\boxed{R_\lambda \equiv \frac{L}{\lambda A}} \tag{2.12}$$

The inverse of the thermal conductive resistance is called the **thermal conductive conductance** $K_\lambda$:

$$K_\lambda = \frac{1}{R_\lambda} \equiv \frac{\lambda A}{L}$$  (2.13)

and $\lambda/L$, the thermal conductive conductance per unit area, is called the *specific thermal conductance or transmittance for the conductive flux*. It is $[R_\lambda] = $ K/W and $[K_\lambda] = $ W/K.

This nomenclature is adequate when more (combined) heat transfer modes exist and need to be assembled for analysis. For example, in case of convection heat transfer, an *average convective coefficient or transmittance* $\overline{h}$ may be applied,[14] with $[h] = $ W/m$^2$K, and the resulting **thermal convective resistance** $R_h$ will be

$$R_h \equiv \frac{1}{\overline{h} A}$$  (2.14)

## 2.2.2  Analogy Between Thermal and Electric Conduction

Equation (2.11), together with definitions by Eqs. (2.12, 2.13), bears similarities with **Ohm's Law**,[15] therefore an **analogy between thermal and electrical conduction** can be derived. Ohm's Law yields, for a conducting medium or conductor:

$$\Delta V = RI$$  (2.15)

*The electric potential tension $\Delta V$ across two points of an electrical conductor is directly proportional to the electrical current $I$ flowing through the conductor between the two points; the constant of proportionality, the electrical resistance $R$, is a characteristic of the conductor.*[16]

---

[14]The nature and significance of the *average* coefficient $\overline{h}$ will be scrutinized in Chap. 4. The definition of average implies that its value is *constant* and *uniform*. In addition to convection heat transfer, radiation heat transfer with the surroundings may well be also present.

[15]As proposed by German physicist G.S. OHM, in the early nineteenth century.

[16]Let us recall here some parameters used in the electrical engineering. First of all, the *electrical power* $P$ is defined by:

$$P = \Delta V I = R I^2$$

The *electrical current* $I$ is measured in ampere (A), a fundamental unit in SI (named after the French mathematician and physicist A.- M.- AMPÈRE, beginning of the nineteenth century). The *electric charge* is measured in coulomb (C), a derived unit in SI (named after the French physicist C.- A.- DE COULOMB, beginning of the eighteenth century) as $1 C = 1 A \times s$. The potential tension or *voltage*, and the electromotive force are measured in volt (V), a derived unit (named after the Italian physicist and chemist A.G.A.A. VOLTA, beginning of the eighteenth century): the voltage between two points of a conductor when a current of 1 A dissipates 1 W between those points: $1 V = 1$ W/A.break The *electrical resistance* $R$ is measured in ohm ($\Omega$), a derived unit: the resistance between two

**Fig. 2.6** Scheme usually employed in electrical engineering to represent the two circuit arrangements

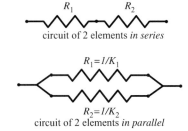

$R_1$     $R_2$

circuit of 2 elements *in series*

$R_1 = 1/K_1$

$R_2 = 1/K_2$

circuit of 2 elements *in parallel*

When the medium is non-homogeneous, a flow network can be cast, so that the current flows through different resistances. The simplest arrangements are the **series** and **parallel**[17] circuits (Fig. 2.6), described by the following:

$$\Delta V = \sum_i R_i I \ , \ \text{for a series circuit;} \quad \Delta V \sum_i K_i = I \ , \ \text{for a parallel circuit}$$

$$(2.16)$$

### 2.2.3 Application of Thermal Circuits: Plane Multilayered Media

In analogy with Eq. (2.15), the temperature difference $\Delta T$ across two points of a thermal conductor is proportional to the thermal power flowing through the conductor surface $A$, featuring a given thermal resistance, $R$. Therefore, for a medium consisting in a variety of plane elements configured **in series** or **in parallel** (as in Fig. 2.7), one has:

$$\dot{Q}_\lambda = \frac{\Delta T}{\sum_i R_i} \ , \ \text{in series;} \quad \dot{Q}_\lambda = \sum_i K_i \, \Delta T \ , \ \text{in parallel} \qquad (2.17)$$

The concepts of composite resistance and conductance are useful in thermal networks, as for simultaneous conduction and convection: in this way, a global exchange coefficient for a composite wall, or **overall heat transfer coefficient** $U$ (i.e., relative to the effective wall surface $A$) can be calculated by knowing the data for each element of the wall, assuming perfect interface contact. From this, the thermal flux can be assessed.

Let us take the example of a container for some substrate assays, used to keep samples at controlled temperature. The reactor wall (Fig. 2.8) consists in a thick

---

points of a conductor when a constant potential difference of 1 V, applied to these points, produces in the conductor a current of 1 A, the conductor not being the seat of any electromotive force: $1 \, \Omega = 1 \, \text{V/A}$.

[17]In accordance with Kirchhoff's Law of current, proposed by German physicist G.R. KIRCHHOFF at mid-nineteenth century.

**Fig. 2.7** Representation of
the two thermal circuit
arrangements: top, *in series*;
bottom, *in parallel*

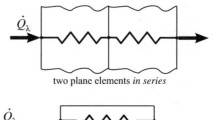

two plane elements *in series*

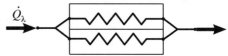

two plane elements *in parallel*

**Fig. 2.8** Heat flowing (in
the direction of decreasing
temperature) through a
composite wall of a
container and related thermal
circuit arrangement.
Internally, air flows across a
number of samples

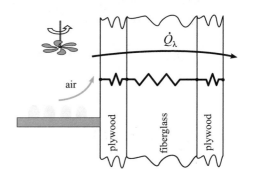

layer of fiberglass, having high insulation property ($\lambda_f = 0.035$ W/mK, thickness
$L_f = 8.0$ cm), within two sheets of structural plywood ($\lambda_p = 0.11$ W/mK, thickness
$L_p = 1.0$ cm). The reactor sits in a room at $T_\infty = 10\,°$C, with an average convective
transmittance $\overline{h}_\infty$ of 5.0 W/m²K. An internal fan is provided to uniform the thermal
regime, with $T_0 = 40\,°$C and $\overline{h}_0 = 20$ W/m²K.

   In order to know how much heat must be provided to maintain the desired regime,
the value of the thermal power $\dot{Q}_\lambda$ flowing through the wall is needed (in the direction
taken by temperature decrease). In this case, we must assume that the its overall
thickness is small compared to its height and width. This corresponds to the circuit
arrangement of Fig. 2.7, top (plane elements in series).

   Recalling Eq. (2.11):

$$\dot{Q}_\lambda \equiv \dot{q}_\lambda A = U A (T_0 - T_\infty) \qquad (2.18)$$

and based on the first of Eq. (2.17), we need to take the sum of the resistances $R_i$, so
that

**Fig. 2.9** Qualitative
temperature distribution
through the composite wall
and at its exposed sides

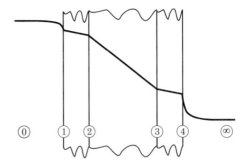

$$\frac{1}{UA} = \frac{1}{\overline{h}_0 A} + \frac{L_i}{\lambda_i A} + \frac{1}{\overline{h}_\infty A} =$$

$$\frac{1}{A}\left(\frac{1}{20} + \frac{0.010}{0.11} + \frac{0.080}{0.035} + \frac{0.010}{0.11} + \frac{1}{5.0}\right) = \frac{2.7}{A}\ \text{K/W}$$

therefore:

$$\dot{Q}_\lambda = \frac{A}{2.7}(40 - 10) = 11 \times A\ \text{W}$$

which can be multiplied by the actual heat transfer surface $A$.

In order to draw in Fig. 2.9 the temperature drop across the wall, we can calculate the temperature value at each side or node (①...④), by using the solving formula Eq. (2.18) for $\dot{q}_\lambda$ *for each circuit segment* at each side or node. To do this, one writes $\dot{q}_\lambda = h_0(T_0 - T_1)$ for the unknown $T_1$, then using the formula again between ① and ② for the unknown $T_2$, and so on resulting with the following values: $T_1 = 39\ °C$, $T_2 = 38\ °C$, $T_3 = 13\ °C$, $T_4 = 12\ °C$. We note then that the larger temperature drop occurs across the fiberglass layer, which features the smaller thermal conductivity.[18]

## 2.3  Fundamental Equation of Thermal Conduction

So far we have dealt with the simplest case of temperature distribution in the steady state through a 1-D geometry by directly using the Fourier's Law of conduction, applied to a plane geometry of finite thickness. To determine the distribution of temperature scalar $T$, for complex geometries and in the unsteady regimes, however, it is necessary to resort to a *generalized formulation*. This formulation will be applied, through the energy balance, and solved by using three different techniques:

---

[18] We also note the typical curvilinear variation of $T$ by the wall sides, due to the *convective boundary layer* which we will deal with when studying Chap. 4: in this case, the temperature drop due to the thermal resistance defined by Eq. (2.14) is smaller where the forced convection is applied (at side $i$), being this mechanism stronger than the external free convection (at side $\infty$) in transferring the heat from/to the side.

**Fig. 2.10** Infinitesimal $CV$
and its confining $CS$ in the
$x - y$ plane, with indication
of a point $P$ and its
coordinates in its reference
frame

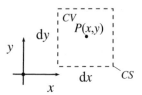

1. formal analytical methods, as in Sects. 2.3.3–2.3.5 (pp. 31–44), for plates and
   cylinders in the steady state;
2. a combination of analytical and graphical tools, as in Sect. 2.4 (p. 49), for plates
   in the unsteady state;
3. numerical methods for any arbitrary geometry and state, as outlined in Sect. 2.5
   (p. 55).

Let us consider first the flow of heat through an infinitesimal $CV$ fixed in a Cartesian (rectangular) 2-D space,[19] and confined by its $CS$, as represented in Fig. 2.10.

### 2.3.1  Equation of Energy Conservation in Rectangular Coordinates

Then, let us apply the energy balance on the $CV$ in terms of *fluxes* (thermal power)
through the $CS$ and in term of *internal energy e* (with $[e]=$ J/kg), work:

$$\boxed{\dot{Q}_\lambda \text{ entering } CS} + \boxed{\text{source/sink of } e \text{ in } CV, \text{ in the unit } \theta}$$

$$= \boxed{\dot{Q}_\lambda \text{ exiting } CS} + \boxed{\text{variation of } e \text{ in } CV, \text{ in the unit } \theta} \qquad (2.19)$$

where the **source/sink term is the volumetric energy flux** $\dot{e}'''$, in unit time and space
(i.e., the $CV$),[20] due to any PCB occurrence of interest.

We had the opportunity already to anticipate, in the formal framework presented first in Sect. 2.1.1 (p. 13), on the *multiphysics nature* of term $\dot{e}'''$ that
accounts **for any internal generative or dissipative effects**. Generally, this volumetric energy flux can be taken as *extrinsic*, as with laser/radiative heat treatment,

---

[19] With this choice of a rectangular reference system, lengths along $z$ can be taken as unitary.

[20] The three primes $'''$ are purposeful to the fact that $\dot{e}'''$ refers to the *unit volume*. Therefore,
$[\dot{e}''']=$ W/m$^3$.

ultrasound/electromagnetic processing, or *intrinsic*, as with phase-change,[21] metabolic effects in living substrates, heat of chemical reaction, radioactive decay.

Here, for sake of simplicity, $\dot{e}'''$ is taken as *uniform across the CV*; note that **it can be positive or negative, depending on the case at stake.** In Eq. (2.19), having written this source term on the left-hand side, when its sign is positive its contribution of energy to the $CV$ is positive, as well.

This balance can be written based on a truncated *Taylor's series expansion.*[22] Equivalently, we can consider that displacing by d$x$ the application point of the subject variable, one has the original quantity plus a variation of the variable, multiplied by the displacement:

$$\dot{Q}_x(x + dx) \equiv \dot{Q}_x(x) + \frac{\partial \dot{Q}_x(x)}{\partial x} dx \qquad (2.20)$$

With this idea in mind, the terms of Eq. (2.19) can be written down as depicted in Fig. 2.11. Then, assuming *constant density and specific heat*, this balance can be written out as follows:

$$\left(\dot{Q}_x + \dot{Q}_y\right) \pm \dot{e}''' dx dy =$$
$$\left(\dot{Q}_x + \frac{\partial \dot{Q}_x}{\partial x} dx + \dot{Q}_y + \frac{\partial \dot{Q}_y}{\partial y} dy\right) + \rho c_p (dx dy) \frac{\partial T}{\partial \theta}$$

or

$$-\left(\frac{\partial \dot{Q}_x}{\partial x} dx + \frac{\partial \dot{Q}_y}{\partial y} dy\right) \pm \dot{e}''' dx dy = \rho c_p (dx dy) \frac{\partial T}{\partial \theta}$$

From now on, a positive sign of source term $\dot{e}'''$ represents a positive contribution of energy to the $CV$.

---

[21] Phase-change-related heat flux, such as for evaporation/condensation, is common in biosubstrate processing, in presence of an internal phase-changing constituent, and deserves a dedicated formulation. As a frequent example, evaporation of liquid water may well occur directly within water-saturated substrates, subject to heat supply. In this case, $\dot{e}'''$ represents the *latent* cooling rate due to evaporation and can be computed as

$$\dot{e}''' = -\dot{m}''' \Delta h_{vap}$$

where $\dot{m}'''$ is the *volumetric flux of water vapor* or *evaporation rate*, in kg/m³s, and $\Delta h_{vap}$ is the *latent heat of evaporation* of water, in kJ/kg.
One such applicative case is presented in Sect. 6.3.

[22] Named after English mathematician B. TAYLOR (early eighteenth century), this series dictates that

$$\dot{Q}_x(x + dx) \equiv \dot{Q}_x(x) + \frac{\partial \dot{Q}_x(x_0)}{\partial x} dx + a \text{ truncation error due to higher-order terms [6]}$$

This *expansion* can be generalized for functions of more variables.

**Fig. 2.11** Infinitesimal $CV$, interested by fluxes through its $CS$ (represented by black line arrows), and source/sink of internal energy right within the same $CV$

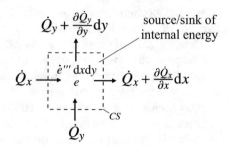

Now, let us plug down the Fourier's Law referred to $x$ or $y$ coordinates:

$$\dot{Q}_x = -\lambda A \frac{dT}{dx} \quad \text{or} \quad \dot{Q}_y = -\lambda A \frac{dT}{dy} \qquad \text{(2.4 revisited)}$$

Multiplying by the unit sides $dx \times 1$ and $dy \times 1$, one has:

$$\lambda \left( \frac{\partial^2 T}{\partial x^2} dxdy + \frac{\partial^2 T}{\partial y^2} dxdy \right) \pm \dot{e}''' dxdy = \rho c_p (dxdy) \frac{\partial T}{\partial \theta}$$

which holds $T$ as the independent variable. Simplifying we have the **fundamental Equation of heat transfer by conduction,**[23] *in dimensional Cartesian form*:

$$\boxed{\rho c_p \frac{\partial T}{\partial \theta} = \lambda \left( \frac{\partial^2 T}{\partial x^2} + \frac{\partial^2 T}{\partial y^2} \right) \pm \dot{e}'''} \qquad (2.21)$$

In a 3-D space, an appropriate third term (the second $z$ derivative of $T$) shows up in the parentheses. *In vector form*[24] we have

---

[23] Also called the *heat equation*.

[24] $\nabla^2$ is the *Laplace* operator (after French physicist, mathematician, astronomer, and statesman P.- S. LAPLACE, early nineteenth century). When applied to a scalar, gives its *Laplacian*. See also Note 13. When using the rectangular 3-D coordinates, the Laplace operator is given by

$$\nabla^2 \equiv \frac{\partial^2}{\partial x^2} + \frac{\partial^2}{\partial y^2} + \frac{\partial^2}{\partial z^2}$$

In cylindrical coordinates, it is

$$\nabla^2 \equiv \frac{\partial^2}{\partial r^2} + \frac{1}{r}\frac{\partial}{\partial r} + \frac{1}{r^2}\frac{\partial^2}{\partial \phi^2} + \frac{\partial^2}{\partial z^2}$$

Finally, in spherical coordinates, it is

$$\nabla^2 \equiv \frac{\partial^2}{\partial r^2} + \frac{2}{r}\frac{\partial}{\partial r} + \frac{1}{r^2 \sin \gamma}\frac{\partial}{\partial \gamma}\left( \sin \gamma \frac{\partial}{\partial \gamma} \right) + \frac{1}{r^2 \sin^2 \gamma}\frac{\partial^2}{\partial \phi^2}.$$

$$\underbrace{\frac{1}{\alpha}\frac{\partial T}{\partial \theta}}_{①} = \underbrace{\nabla^2 T}_{②} \pm \underbrace{\frac{\dot{e}'''}{\lambda}}_{③} \qquad (2.22)$$

The various terms in Eq. (2.22) all represent thermal fluxes in the unit volume, with dimensions $(K/m^2)$, *in competition* in every medium point: term ① represents the (transient) *thermal inertia*, term ② the *conduction or heat diffusion flux*, term ③ the *internal source/sink term*. The positive sign of term ③ pertains to the heat generation.

Equation (2.22) is a second-order, time-*parabolic*, and space-*elliptic* PDE. This problem is also called a "Marching" problem, as the solution of the dependent function (temperature) must be computed by proceeding in time from the initial state, while satisfying the conditions at the boundaries.[25]

*Equation* (2.22) together with an **initial condition** (stating the temperature distribution in the $CV$ at the beginning of the process), and with the appropriate number of **boundary conditions** (BCs) onto the $CS$, *allows one to determine the function* $T(x, y, \theta)$, in the specified assumptions of solid medium or liquid medium at rest, isotropic with respect to $\lambda$, and with $\rho$ and $c$ constant with time. Four different kinds of BC are employed in heat transfer:

1. when the temperature itself is specified, we have the first-kind BC, or *Dirichlet boundary condition*;[26]
2. when the temperature slope or gradient (thermal flux) is specified, we have the second-kind BC, or *Neumann boundary condition*.[27] When nonzero, the *sign* of this quantity (the direction of thermal flux with respect to the $CV$ at stake) *is always opposite of the sign of the temperature gradient*;
3. when a linear combination of these two cases is specified, we have the third-kind BC, or *Robin boundary condition*;[28]
4. a fourth-kind BC can also be cast, involving *radiative heat transfer.*[29]

Some of these BCs will be illustrated in the following Sects. 2.3.3 (p. 31) and 2.3.5 (p. 44). BCs may well include any multiphysics effect of sort, such as phase-change.

---

[25] PDEs can be usefully classified in *parabolic*, *hyperbolic*, and *elliptic* types, in analogy with second-order equations appearing in analytic geometry [3]. In addition to the parabolic type, an example of hyperbolic PDE is the *wave equation* used for the description of waves such as sound, light, and in liquid at interface

$$\frac{\partial^2 u}{\partial \theta^2} = c^2 \nabla^2 u$$

with $u$ the subject scalar (e.g., the wave's mechanical displacement) and $c$ the propagation speed. Elliptic PDEs are typically used in equilibrium phenomena, such the one that would result in the steady state, when dropping the transient term ③ in the heat equation, Eq. (2.22).

[26] After German mathematician P.G.L. LEJEUNE DIRICHLET (mid-nineteenth century).

[27] After German mathematician C. NEUMANN (late nineteenth century).

[28] After French mathematician V.G. ROBIN (late nineteenth century).

[29] This nomenclature stems from the *Stefan–Boltzmann Law*

$$E_b = \sigma T^4$$

An appropriate allocation of corresponding BCs will be at stake for every transport phenomena mechanism in this book.[30]

### 2.3.2 Other Coordinate Systems

Coordinate systems other than the rectangular one can be employed if needed [4]. With a *cylindrical coordinate system* ($r$, $\phi$, $z$, Figure at left in Note 12), we can operate in the same way as we had so far, using for an infinitesimal $CV$ the conductive thermal fluxes[31] that cross the various $CS$s:

$$\dot{q}_{\lambda r} \equiv -\lambda \frac{\partial T}{\partial r}; \quad \dot{q}_{\lambda \phi} \equiv -\frac{\lambda}{r} \frac{\partial T}{\partial \phi}; \quad \dot{q}_{\lambda z} \equiv -\lambda \frac{\partial T}{\partial z} \tag{2.23}$$

With these definitions, the **fundamental Equation of heat transfer by conduction in cylindrical coordinates** with constant and uniform properties and internal source/sink can be written as follows:

$$\boxed{\frac{1}{\alpha} \frac{\partial T}{\partial \theta} = \frac{1}{r} \frac{\partial}{\partial r} \left( r \frac{\partial T}{\partial r} \right) + \frac{1}{r^2} \frac{\partial^2 T}{\partial \phi^2} + \frac{\partial^2 T}{\partial z^2} \pm \frac{\dot{e}'''}{\lambda}} \tag{2.24}$$

Similarly, we can obtain the Equation in a *spherical coordinates system* ($r$, $\gamma$, $\phi$, Figure at right in Note 12):

$$\dot{q}_{\lambda r} \equiv -\lambda \frac{\partial T}{\partial r}; \quad \dot{q}_{\lambda \gamma} \equiv -\frac{\lambda}{r} \frac{\partial T}{\partial \gamma}; \quad \dot{q}_{\lambda \phi} \equiv -\frac{\lambda}{r \sin \gamma} \frac{\partial T}{\partial \phi} \tag{2.25}$$

---

which describes the *power radiated* $E_b$ from a so-called *black body* based on its absolute temperature to the fourth power, $\sigma$ being a constant. This Law was formulated jointly by the Slovenian-Austrian mathematician and physicist J. STEFAN and the Austrian physicist L. BOLTZMANN (late nineteenth century).

[30] All of the governing equations in this book are elliptic in space and parabolic in time. For space-elliptic PDEs, the second space derivatives (term ② in Eq. (2.22)) require two boundary conditions to be assigned in each spatial coordinate. In principle, this could be done by prescribing conditions at two locations or by prescribing two conditions at one location: only the former is appropriate. For example, prescribing both the function and its derivative at the same location, i.e., the so-called *Cauchy boundary condition* (after French mathematician A.- L. CAUCHY), leads to BCs overspecification and improperly-posed problems. See discussion in [Morse, P.M., Feshbach, H.: Methods of Theoretical Physics. McGraw-Hill, New York (1953) p. 690] and in [Jaluria, Y., Torrance, K. E.: Computational Heat Transfer. Hemisphere Publishing, Washington (1986) p. 18].

[31] Note that, due to the nature of angular coordinate $\phi$, all three fluxes are dimensionally homogeneous.

With these definitions, the **fundamental Equation of heat transfer by conduction in spherical coordinates** with constant and uniform properties and internal source/sink can be written as follows:

$$\frac{1}{\alpha}\frac{\partial T}{\partial \theta} = \frac{1}{r^2}\frac{\partial}{\partial r}\left(r^2 \frac{\partial T}{\partial r}\right) + \frac{1}{r^2 \sin \gamma}\frac{\partial}{\partial \gamma}\left(\sin \gamma \frac{\partial T}{\partial \gamma}\right)$$
$$+ \frac{1}{r^2 \sin^2 \gamma}\frac{\partial}{\partial \phi}\left(\frac{\partial T}{\partial \phi}\right) \pm \frac{\dot{e}'''}{\lambda} \tag{2.26}$$

### 2.3.3 Temperature Distribution in the Steady State, 1-D

Now we can develop the solutions for some basic cases and geometries.

#### 2.3.3.1 Rectangular Coordinates, No Source

Let us turn again to

$$\frac{1}{\alpha}\frac{\partial T}{\partial \theta} = \nabla^2 T \pm \frac{\dot{e}'''}{\lambda} \tag{2.22 revisited}$$

for the infinite plate in Fig. 2.12 in the steady state and with *no volumetric source/sink flux*:[32]

$$\frac{d^2 T}{dx^2} = 0 \tag{2.27}$$

The BCs of this problem are of the first-kind:

$$T = T_0, \text{ for } x = 0; \quad T = T_L, \text{ for } x = L \tag{2.28}$$

**Fig. 2.12** *CV* for the plate (lengths along *y* and *z* are taken as unitary), with indication of limiting temperatures and thickness

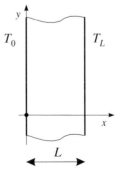

---

[32] Also called the *Laplace's equation*.

Integrating Eq. (2.27) twice one has:

$$T = ax + b \tag{2.29}$$

Applying the two BCs:

$$T = T_0 + (T_L - T_0)\,\frac{x}{L} \tag{2.30}$$

As proof, Eq. (2.30) can be derived and, when substituting the Fourier's Law in the differential form

$$\dot{q}_{\lambda y} = -\lambda \frac{\partial T}{\partial y} \tag{2.6}$$

one has again:

$$\dot{q}_{\lambda} = \frac{\lambda}{L}\,(T_0 - T_L)$$

with the bulk thermal power crossing the plate denoted with $\dot{Q}_h$ (Fig. 2.13). It will be simply $\dot{Q}_h = \dot{q}_{\lambda} \times 1$, considering the unitary surface affected. We see that, if $T_0 > T_L$, the thermal flux $\dot{q}_{\lambda}$ is directed toward the positive $xs$ ; $\dot{q}_{\lambda}$ is conserved during the flow of heat from $x = 0$ to $x = L$; the isotherms are equidistant and parallel.

#### 2.3.3.2 Rectangular Coordinates, Effect of Internal Source

The governing Equation

$$\frac{1}{\alpha}\frac{\partial T}{\partial \theta} = \nabla^2 T \pm \frac{\dot{e}'''}{\lambda} \tag{2.22 revisited}$$

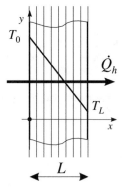

**Fig. 2.13** Distribution of temperature for the plate in the steady state, with no internal source/sink, with superposition of some isotherms parallel to the plate's sides and indication of the bulk thermal power $\dot{Q}_h$ removed by convection from the plate

for the infinite plate of Fig. 2.12 in the steady state but *with a volumetric generation flux and a heat removal by an air draft from the sides*,[33] gives this time:

$$\frac{d^2 T}{dx^2} + \frac{\dot{e}'''}{\lambda} = 0 \qquad (2.31)$$

The distribution of temperature will be stationary as the heat due to the volumetric flux $\dot{e}'''$ will be removed from the sides by a *convective flux* $\dot{q}_h$ (Fig. 2.14). Due to the *energy conservation at the sides*, the BCs therefore are of the second-kind:

$$-\dot{q}_\lambda = \dot{q}_h = \overline{h}\,(T - T_\infty)\,, \ \text{for } x = -L; \quad \dot{q}_\lambda = \dot{q}_h = \overline{h}\,(T - T_\infty)\,, \ \text{for } x = L \qquad (2.32)$$

where $\overline{h}$ is an average convective transmittance. To substitute the thermal conductive flux $\dot{q}_\lambda$, the Fourier's Law

$$\dot{q}_\lambda = -\lambda \frac{\partial T}{\partial y} \qquad (2.6 \text{ revisited})$$

must be invoked to each side:

$$\lambda \frac{dT}{dx} = \overline{h}\,(T - T_\infty)\,, \ \text{for } x = -L \qquad (2.33a)$$

$$-\lambda \frac{dT}{dx} = \overline{h}\,(T - T_\infty)\,, \ \text{for } x = L \qquad (2.33b)$$

It is important to note that, with these assumptions, *both geometry and BCs are symmetrical with respect to the center line* for $x = 0$. This yields that the *distribution of temperature must be symmetrical* as well, about such line; that is,

$$\frac{dT}{dx} = 0, \ \text{for } x = 0 \qquad (2.34)$$

**Fig. 2.14** A symmetrical version of the plate $CV$, with indication of thickness and removal of heat by the sides

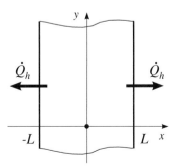

---

[33] Also called the *Poisson's equation*, after French mathematician and physicist S.D. POISSON (early nineteenth century).

Then, the center line corresponds to an insulation boundary, and we can focus on the right $CV$'s half; the BCs become:

$$\frac{dT}{dx} = 0, \text{ for } x = 0; \quad -\lambda \frac{dT}{dx} = \bar{h}\,(T - T_\infty), \text{ for } x = L \qquad (2.35)$$

To carry out the integration, let us separate the differential elements in the denominator of at the left-hand side of Eq. (2.31):

$$\frac{d^2 T}{dx} = -\frac{\dot{e}'''}{\lambda}dx$$

Integrating both-hand sides, one has:

$$\frac{dT}{dx} = -\frac{\dot{e}'''}{\lambda}x + a \qquad (2.36)$$

where a first constant $a$ shows up. Separating the differential terms a second time:

$$dT = -\frac{\dot{e}'''}{\lambda}x dx + a dx$$

Now integrating again on both sides yields the general integral, which is also affected by a second constants $b$:

$$T(x) = -\frac{\dot{e}'''}{\lambda}\frac{x^2}{2} + ax + b \qquad (2.37)$$

The BCs Eq. (2.35) can be enforced, one at a time:

$$\text{for } x = 0 \Rightarrow \frac{dT}{dx} = -\frac{\dot{e}'''}{\lambda}x + a = 0 \Rightarrow a = 0 \qquad (2.38)$$

The first constant has been determined. Then again:

$$\text{for } x = L \Rightarrow T(x) = -\frac{\dot{e}'''}{\lambda}\frac{x^2}{2} + b \Rightarrow T(L) = -\frac{\dot{e}'''}{\lambda}\frac{L^2}{2} + b \qquad (2.39)$$

This last result is an expression for $T(L)$, which can be substituted in the second BC. Thus

$$-\lambda\frac{dT}{dx} = -\frac{\dot{e}'''}{\lambda}\bar{h}\frac{L^2}{2} + \bar{h}b - \bar{h}T_\infty \qquad (2.40)$$

Now let us note that, when writing Eq. (2.36) for $x = L$, we had another relationship between the conductive flux and the generated heat already. Hence,

$$\dot{e}''' L = -\lambda \frac{dT}{dx} \tag{2.41}$$

Equating with the right-hand side of Eq. (2.40), one has:

$$\dot{e}''' L = -\frac{\dot{e}'''}{\lambda} \bar{h} \frac{L^2}{2} + \bar{h} b - \bar{h} T_\infty$$

or

$$b = -\frac{\dot{e}''' L}{\bar{h}} + \frac{\dot{e}''' L^2}{2\lambda} + T_\infty \tag{2.42}$$

The second constant has been determined. We can then obtain the solution for the temperature and its maximum value. Plugging the values for $a$ and $b$ in the general integral of Eq. (2.37), the solution for $T$ becomes:

$$T(x) = -\frac{\dot{e}'''}{\lambda} \frac{x^2}{2} + \frac{\dot{e}''' L^2}{2\lambda} + \frac{\dot{e}''' L}{\bar{h}} + T_\infty$$

The first two terms on the right-hand side can be multiplied and divided for $L^2$:

$$T(x) - T_\infty = \frac{\dot{e}''' L^2}{2\lambda}\left(1 - \frac{x^2}{L^2}\right) + \frac{\dot{e}''' L}{\bar{h}}$$

or

$$T(x) - T_\infty = \frac{\dot{e}''' L^2}{2\lambda}\left[1 - \left(\frac{x}{L}\right)^2\right] + \frac{\dot{e}''' L}{\bar{h}} \tag{2.43}$$

To calculate the removed convective power $\dot{Q}_h$, it is sufficient to note that if we plug here $x = L$ and $T = T_L$, based on the second BC we find again the initial assumption of $\dot{Q}_h = \dot{e}''' L$, considering the unitary sizes along $y$ and $z$.

The maximum temperature is found on the center line:

$$T_{max} - T_\infty = T(x)|_{x=0} - T_\infty = \frac{\dot{e}''' L^2}{2\lambda} + \frac{\dot{e}''' L}{\bar{h}} = \frac{\dot{e}''' L^2}{2\lambda}\left(1 + \frac{2\lambda}{\bar{h} L}\right) \tag{2.44}$$

and the temperature distribution is a *parabola* (Fig. 2.15): notice again the typical curvilinear temperature progress that would result, if we were to solve for the temperature distribution in the adjacent fluid, as well. We will postpone and focus on this part in Chap. 4.

**Fig. 2.15** Distribution of
temperature for the plate in
the steady state, with the
effect of internal source/sink

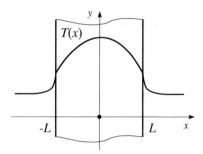

Equation (2.44) can also written as:

$$T_{\max} = \frac{\dot{e}''' L^2}{2\lambda} \left(1 + \frac{1}{\mathrm{Bi}}\right) \quad \text{with} \quad \boxed{\mathrm{Bi} \equiv \frac{\overline{h} L}{\lambda}} \tag{2.45}$$

The **Biot number** Bi, a dimensionless parameter of thermal transport, informs on the *competition between external convection and internal conduction*.[34] In other words, this parameter determines whether or not the temperatures inside a body will vary significantly in space, when subject to convection.

Some algebraic manipulation will clarify on the Bi number meaning:

$$\mathrm{Bi} \equiv \frac{\overline{h} A}{\lambda A / L} = \frac{\text{superficial conductance}}{\text{internal conductive conductance}}$$
$$= \frac{\text{internal conductive resistance}}{\text{superficial resistance}} \tag{2.46}$$

There are two ranges worth noting:

1. Bi ≫ 1—good thermal contact between fluid and solid: the temperature at the interface is very close to $T_\infty$, and the temperature inside the body varies considerably
2. Bi ≪ 1—bad thermal contact, the interface temperature is very close to the maximum one (on the center line), and the temperature is uniform inside the body

Naturally, Bi numbers can be used and are encountered for other geometries, as well.

As a general notation, $L$ in the definition of Eq. (2.46) can be regarded as a *reference length* $V/A$, obtained by the capacity or volume $V$ of the $CV$ and the extension or surface area $A$ of the $CS$.

---

[34] As proposed by French physicist J.- B. BIOT, in the early nineteenth century.

### 2.3.3.3 Cylindrical Coordinates, Effect of Internal Source

The governing Equation is

$$\frac{1}{\alpha}\frac{\partial T}{\partial \theta} = \frac{1}{r}\frac{\partial}{\partial r}\left(r\frac{\partial T}{\partial r}\right) + \frac{1}{r^2}\frac{\partial^2 T}{\partial \phi^2} + \frac{\partial^2 T}{\partial z^2} \pm \frac{\dot{e}'''}{\lambda} \qquad (2.24)$$

can be recalled, with reference to an infinite cylinder with BCs uniform along $\phi$,[35] *with a volumetric generation flux and convective heat transfer from the side.* Such a situation can be found, for example, when configuring an electric heating resistor[36] for an oven. Similarly than for Eq. (2.31), we end up with:

$$\frac{1}{r}\frac{d}{dr}\left(r\frac{dT}{dr}\right) + \frac{\dot{e}'''}{\lambda} = 0 \qquad (2.47)$$

The assumptions of the cylindrical geometry and boundary conditions of constant and uniform temperature $T_\infty$ and average convective transmittance $\overline{h}$ ensure that the problem is symmetrical with respect to the axis for $r = 0$, and therefore, the geometry of Fig. 2.16 can be considered. The distribution of temperature will be stationary as the heat due to the volumetric flux $\dot{e}'''$ will be removed from the side by a *convective flux* $\dot{q}_h$. The BCs then become:

$$\frac{dT}{dr} = 0, \text{ for } r = 0; \quad -\lambda\frac{dT}{dr} = \overline{h}\left(T - T_\infty\right), \text{ for } r = R \qquad (2.48)$$

**Fig. 2.16** Symmetrical geometry for a cylinder, with indication of thickness and removal of heat by the side

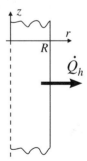

---

[35]The first assumption is valid, that is, neglecting the effect of extremal sides along $z$ (Figure at left in Note 12) when the height $L$ is at least one OoM greater than the radius $R$. With the second assumption, this case is called the *circular* coordinates geometry.

[36]If the material has electrical conductivity $\lambda_e$ ($1/\Omega\text{cm}$), and is subject to a current density $I$ ($\text{A/cm}^2$), the heat generation due to the Joule heating effect (after English physicist J.P. JOULE, at mid-nineteenth century) is $\dot{e}''' = I^2/\lambda_e$ ($\Omega\text{A}^2/\text{cm}^3$).

To carry out the integration, let us separate the differential elements of Eq. (2.47) one first time:

$$d\left(r\frac{dT}{dr}\right) = -\frac{\dot{e}'''}{\lambda}r\,dr$$

Integrating both-hand sides and dividing by $r$, one has:

$$\frac{dT}{dr} = -\frac{\dot{e}'''}{2\lambda}r + \frac{a}{r}$$

Let us use the BC at the axis. To avoid the inconsistency of $dT/dr \rightarrow \infty$ for $r = 0$, we soon realize that $a$ must be equal to 0:

$$\frac{dT}{dr} = -\frac{\dot{e}'''}{2\lambda}r \qquad (2.49)$$

The first constant has been determined. Then, integrating the Eq. (2.49) again, we come up with the general integral:

$$T(r) = -\frac{\dot{e}'''}{4\lambda}r^2 + b \qquad (2.50)$$

This can be written for $r = R$:

$$T_R + \frac{\dot{e}'''}{4\lambda}R^2 = b \qquad (2.51)$$

In order to compute for $b$, the actual value of $T_R$ still needs to be determined. To do this, let us use the derivative of $T$, Eq. (2.49), by plugging it in the second BC. One has:

$$\frac{dT}{dr} = -\frac{\dot{e}'''}{2\lambda}r = -\frac{\overline{h}}{\lambda}\left(T - T_\infty\right), \text{ for } r = R$$

or

$$\lambda\frac{\dot{e}'''}{2\lambda}R = \overline{h}\left(T_R - T_\infty\right) \qquad (2.52)$$

From this, the value of $T_R$ is found:

$$T_R = T_\infty + \frac{\dot{e}'''}{2\overline{h}}R$$

which can be substituted in Eq. (2.51) to yield

$$b = T_\infty + \frac{\dot{e}'''}{2\overline{h}}R + \frac{\dot{e}'''}{4\lambda}R^2 \qquad (2.53)$$

The second constant has been determined. Then plugging the value for $b$ in the general integral of Eq. (2.50), the solution for $T$ becomes:

$$T(r) = \frac{\dot{e}'''}{2\bar{h}}R + \frac{\dot{e}'''R^2}{\lambda}\left[1 - \left(\frac{r}{R}\right)^2\right] + T_\infty$$

or

$$T(r) = T_R + \frac{\dot{e}'''}{4\lambda}R^2\left[1 - \left(\frac{r}{R}\right)^2\right] \tag{2.54}$$

To calculate the removed power $\dot{Q}_h$, it is sufficient to note that this time the area $A$ crossed by the heat flux, for the entire cylinder length $L$, depends on the arbitrary $r$: $A(r) \equiv 2\pi r L$. As we write Fourier's Law, or Eq. (2.4), in terms of $r$:

$$\dot{Q}_\lambda = -\lambda A(r)\frac{\mathrm{d}T}{\mathrm{d}r} \tag{2.55}$$

we may substitute Eq. (2.54) in it (knowing its derivative, or Eq. (2.49)), to find again the initial assumption based on the thermal balance at $R$:

$$\dot{Q}_h = \dot{Q}_\lambda\big|_R = \dot{e}'''\pi R^2 L \tag{2.56}$$

The maximum temperature is found on the center line for $r = 0$:

$$T_{\max} = T_R + \frac{\dot{e}'''}{4\lambda}R^2 \tag{2.57}$$

and the temperature distribution is a *paraboloid* (with the chosen geometry representation), which is the surface created from the rotation of a parabola about its axis (Fig. 2.17).

**Fig. 2.17** Distribution of temperature for the cylinder in the steady state, with the effect of internal source/sink

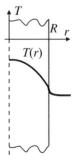

### 2.3.3.4  Cylindrical Shells

Frequently, the temperature distribution and the thermal flux across a infinite cylinder shell (tube), without source/sink term, are of interest. As in Sect. 2.2.3 (p. 23), the overall heat transfer coefficient $U$ can be calculated and, from this, the thermal flux can be assessed. With the same assumptions than the preceding Section, the temperature distribution depends only on $r$. Let us start with recalling the governing Equation again

$$\frac{1}{\alpha}\frac{\partial T}{\partial \theta} = \frac{1}{r}\frac{\partial}{\partial r}\left(r\frac{\partial T}{\partial r}\right) + \frac{1}{r^2}\frac{\partial^2 T}{\partial \phi^2} + \frac{\partial^2 T}{\partial z^2} \pm \frac{\dot{e}'''}{\lambda} \tag{2.24}$$

pretending that heat is conducted in the $r$-direction:

$$\frac{d}{dr}\left(r\frac{dT}{dr}\right) = 0 \tag{2.58}$$

Let us assume first a *monolayer tube*, with *specified inner temperature and outer heat transfer*. A portion of the tube section is depicted in Fig. 2.18 which includes the related thermal circuit (as in Fig. 2.6, top). The boundary conditions are translated into the following:

$$T(r) = T_1, \text{ for } r = r_1, \quad -\lambda\frac{dT}{dr} = \overline{h}_\infty (T - T_\infty), \text{ for } r = r_2 \tag{2.59}$$

Similarly to the integration of Eq. (2.47), one has

$$r\frac{dT}{dr} = a \tag{2.60}$$

and, separating variables and integrating again, we have the general integral

$$T(r) = a\ln r + b \tag{2.61}$$

The BCs give

$$T_1 = a\ln r_1 + b, \quad -\lambda\frac{a}{r_2} = \overline{h}_\infty (a\ln r_2 + b - T_\infty)$$

Solving for $a$ and $b$, we obtain

$$a = \frac{(T_\infty - T_1)}{\ln\frac{r_2}{r_1} + \frac{\lambda}{\overline{h}_\infty r_2}}, \quad b = T_1 - \frac{(T_\infty - T_1)}{\ln\frac{r_2}{r_1} + \frac{\lambda}{\overline{h}_\infty r_2}}\ln r_1$$

**Fig. 2.18** *CV* portion of the monolayer tube, with representation of the thermal circuit

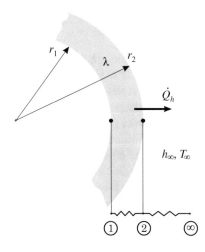

and the solution Eq. (2.61) becomes[37]

$$T(r) = T_1 - \frac{(T_1 - T_\infty)}{\ln \frac{r_2}{r_1} + \frac{\lambda}{\bar{h}_\infty r_2}} \ln \frac{r}{r_1} \tag{2.62}$$

This is a logarithmic distribution that we could report by using the same symmetrical geometry of Fig. 2.16.

To calculate the removed power $\dot{Q}_h$, we simply recall the overall heat transfer coefficient $U$ definition

$$\dot{Q}_\lambda \equiv U A (T_0 - T_\infty) \tag{2.18 revisited}$$

and write

$$\dot{Q}_h = \frac{T_1 - T_\infty}{1/UA} \tag{2.63}$$

From examining the thermal circuit in Fig. 2.18, we proceed by summing the resistances in the two-segment circuit

$$R_\lambda \equiv \frac{L}{\lambda A} \tag{2.12}$$

$$R_h \equiv \frac{1}{\bar{h} A} \tag{2.14}$$

---

[37]It is easy to verify that, in case of two Dirichlet BCs in Eq. (2.59), the solution is simply

$$T(r) = T_1 - \frac{(T_1 - T_\infty)}{\ln \frac{r_2}{r_1}} \ln \frac{r}{r_1}.$$

so that

$$\frac{1}{UA} = \frac{\ln \frac{r_2}{r_1}}{2\pi \lambda L} + \frac{1}{\overline{h}_\infty 2\pi r_2 L} \tag{2.64}$$

The area $A$ need not be specified since all we need is the $UA$ product. However, often a value of $U$ will be quoted based on either the inside or outside area; then the appropriate $A$ value must be used in Eqs. (2.63) and (2.64). Indeed, notice that in contrast to the plane wall cases seen so far, $A$ for the heat transfer may be different on each side of the tube.

Then we can extend these results, without further analysis, to a *multilayer tube*, with *specified inner and outer heat transfer*. A portion of the tube section is depicted in Fig. 2.19 which includes the related thermal circuit.

To calculate the removed power $\dot{Q}_h$, we write again Eq. (2.63) between the fluid at $T_0$ and the fluid at $T_\infty$:

$$\dot{Q}_h \equiv UA(T_0 - T_\infty) = \frac{T_0 - T_\infty}{1/UA} \tag{2.65}$$

Then we proceed by summing the resistances in the four-segment circuit:

$$\frac{1}{UA} = \frac{1}{\overline{h}_0 2\pi r_1 L} + \frac{\ln \frac{r_2}{r_1}}{2\pi \lambda_A L} + \frac{\ln \frac{r_3}{r_2}}{2\pi \lambda_B L} + \frac{1}{\overline{h}_\infty 2\pi r_3 L} \tag{2.66}$$

**Fig. 2.19** $CV$ of the multilayer tube (consisting in materials A and B), with representation of the thermal circuit

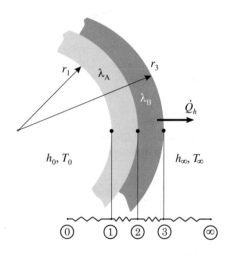

### 2.3.4  Phase Interaction and Discontinuity in Conduction

When the $CV$ at stake comprises more than a material or phase, being subdivided in given sub-$CV$s, one or more interface surface exist. In the spirit of the recursive learning approach proposed in this book, as explained earlier with Fig. 1.1, specific Sections reporting on the inherent basic interaction effects are presented for each transport mechanism. Nowadays, methods of description of transfer phenomena are frequently adopted that deal with different mechanisms at once, due to the mutual interdependence, and often the modeling is extended to the entire $CV$ to avoid segregated solutions needing proper interlacing and additional efforts.

When dealing with conduction heat transfer, only few words are necessary, in preparation of more complex interaction effects that will materialize later. When considering the fundamental Equation of heat transfer by conduction with the associated BC conditions, as explained in Sect. 2.3 (p. 25), an additional BC arises that was implied when describing multilayered shells such as in Sects. 2.2.3 (p. 23) and 2.3.3.4 (p. 40). When two materials A and B with different thermal conductivities are coupled, with assumed perfect interface contact as illustrated in Fig. 2.20, the two temperatures $T_A$ and $T_B$ and the two thermal fluxes must be the same at the interface:

$$T_A(y) = T_B(y), \text{ for } x = 0 \tag{2.67a}$$

$$-\lambda_A \left. \frac{\partial T_A}{\partial x} \right|_{x^-} = -\lambda_B \left. \frac{\partial T_B}{\partial x} \right|_{x^+}, \text{ for } x = 0 \tag{2.67b}$$

As it is evident from examining the temperature discontinuities across different material interfaces in Fig. 2.9, the requirement of Eq. (2.67b) is somehow stronger implying a calculation of the derivative (as will be evidenced in a later Section), and this may pose some calculation difficulty.

The subject of correct consideration of temperature field solution across some interface will be examined in convection heat transfer, where the interface poses more stringent difficulties. Here, it will suffice to carry out the solution of one temperature only, across the interface at $x = 0$, leaving to an appropriate *discretization refinement* or *numerical space grid* (as will be described soon in the next Section), the task of resolving the inherent temperature gradient across the A-B interface.

**Fig. 2.20**  Two different materials sharing one side in perfect thermal contact, with indication of distribution of temperatures and thermal fluxes

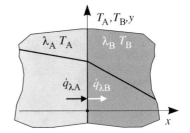

### 2.3.5 Temperature Distribution in the Unsteady State

The unsteady or transient conduction is relevant in many practical cases, such as the thermal treatments to a variety of media. Let us take again the governing Equation

$$\rho c_p \frac{\partial T}{\partial \theta} = \lambda \left( \frac{\partial^2 T}{\partial x^2} + \frac{\partial^2 T}{\partial y^2} \right) \pm \dot{e}''' \tag{2.21}$$

where the source/sink term $\dot{e}'''$ can be left out of the discussion, for sake of simplicity:

$$\frac{\partial^2 T}{\partial x^2} = \frac{1}{\alpha} \frac{\partial T}{\partial \theta} \tag{2.68}$$

The problem consists in determining the temperature distribution and progress in space and time for a generic body (described by its properties and initial thermal state), immersed in a medium and *subject to a sudden thermal perturbation*. This situation can be described by the following body's features, regarding its state and relationship with the environment (Fig. 2.21):

- characteristic semi-length $\Delta x$, volume $V$, and external surface $A$;
- thermal conductivity $\lambda$, density $\rho$, and specific heat $c_p$;
- the body is initially at the uniform temperature $T_i$, while its surface temperature is suddenly brought to $T_0$;
- the body is immersed in a fluid of temperature $T_\infty$, with an average convective transmittance $\bar{h}$.

Let us suppose, for example, that $T_\infty < T_i$; starting from the beginning of the perturbation, the external layer of thickness $\delta x$ cools down and grows with time, its temperature varying from $T_0$ at the external surface, up to the internal $T_i$ (Fig. 2.22). Eventually, after a sufficiently long time, the *thermal perturbation will completely penetrate the body*, which will then be at a constant and uniform $T_\infty$.

It is useful then to define the following four different situations [4]:

1. **at the beginning of the process**, the body temperature starts to be function of time $\theta$ and spatial coordinate $x$: $T = T(x, \theta)$
2. **in a transitional phase**, the thermal perturbation will interest all of the body

**Fig. 2.21** A generic body, immersed in a fluid, whose thermal evolution is a stake. The thermodynamic, geometric, and ambient properties are indicated

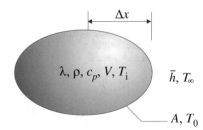

**Fig. 2.22** A body slice along $x$, and the distribution of temperature under heating

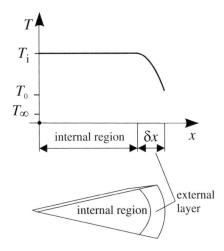

3. **after some time**, the entire body will feature one single temperature, varying with time: $T = T(\theta)$
4. **after a long time**, the body will end in the thermal equilibrium with the surrounding medium.

These first three statements correspond to the following three models, Fig. 2.23:

1. the **initial regime**, or *semi-infinite plate model*: the temperature changes do not penetrate far enough into the body, and the process is confined to its external layer
2. a **general unsteady regime**, or *full model* of conduction in the unsteady state, 1-D, represented by Eq. (2.68) with its ancillary boundary and initial conditions
3. the **late regime**, or *lumped model*: the temperature changes have decayed, and a single value of $T$ for the entire body can be monitored in time.

**Fig. 2.23** Four unsteady-state models, with the temperature distribution along $x$, transiting from the initial to the final regime, with the Marching of time $\theta$

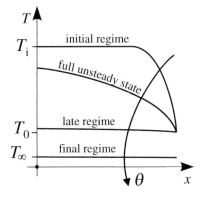

Let us dig first into the first and third models. The second model, the **general unsteady regime**, will be dealt with the general dimensional analysis and solution tools in Sect. 2.4 (p. 49).

### 2.3.5.1 The Initial Regime

When the semi-infinite plate model is applicable (such as in the external layer of a large, poorly conductive body exposed to a weak thermal resistance at its exposed surface, Fig. 2.24), the process is driven by the internal thermal resistance, and the transient temperature distribution can be found by applying a specific analytical technique, called *similarity*.[38] With the development of the technique [4] laying beyond the scope of this book, as well as every possible case, we report here on the analytical solutions for two common BCs applied at body's exposed surface:

1. first-kind (constant surface temperature) $T = T_\infty$, for $x = 0$
2. second-kind (constant convective flux, released from the fluid)
   $\bar{h}(T_\infty - T) = -\lambda(\partial T/\partial x)$, for $x = 0$.

The analytical solution for this problem takes advantage of the special *error* and *complementary error functions*, erf($x$) and erfc($x$), respectively.[39] If Eq. (2.68) is

**Fig. 2.24** Semi-infinite plate geometry and the distribution of temperature under heating

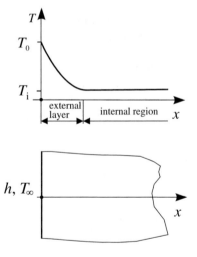

---

[38]Practically, it is seen that all the possible $T = T(x, \theta)$ profiles are similar, that is, have the same shape.

[39]erf and erfc have a sigmoidal shape, as explained in Abramowitz, M., Stegun, I.A.: Handbook of mathematical functions: with formulas, graphs, and mathematical tables. Dover Publications, Mineola (1965). Thus

$$\text{erf}(x) \equiv \frac{2}{\sqrt{\pi}} \int_0^x \exp\left(-t^2\right) \, dt \quad \text{and} \quad \text{erfc}(x) \equiv 1 - \text{erf}(x) = \frac{2}{\sqrt{\pi}} \int_x^\infty \exp\left(-t^2\right) \, dt$$

considered with the initial condition $T = T_i$, for $\theta = 0$ and with a first boundary condition such that $T \to T_i$, for $x \to \infty$, the following analytical solutions can be cast:

1. first-kind BC:

$$\frac{T(x,\theta) - T_\infty}{T_i - T_\infty} = \text{erf}\left[\frac{x}{2\sqrt{\alpha\theta}}\right] \qquad (2.69)$$

2. second-kind BC:

$$\frac{T(x,\theta) - T_\infty}{T_i - T_\infty} = \text{erf}\left[\frac{x}{2\sqrt{\alpha\theta}}\right] +$$
$$\exp\left(\frac{\overline{h}x}{\lambda} + \frac{\overline{h}^2\alpha\theta}{\lambda^2}\right) \text{erfc}\left[\frac{x}{2\sqrt{\alpha\theta}} + \frac{\overline{h}}{\lambda}\sqrt{\alpha\theta}\right] \qquad (2.70)$$

The group $\sqrt{\alpha\theta}$ is the *penetration distance* of the conduction perturbation: an important length scale for the unsteady conduction process. Actually, the semi-infinite plate model is indeed an adequate framework for more complicated body shapes at short times, when this penetration distance is shorter than the transversal dimension of the body (normal to the propagation direction). At longer times instead, when $\sqrt{\alpha\theta}$ is comparable with, or larger than, this transversal dimension (e.g., the radius of the sphere or the cylinder normal to the propagation direction), the conduction distribution becomes multidimensional and a more general method like the one introduced in Sect. 2.5.4 (p. 63) must be invoked.

### 2.3.5.2 The Late Regime

When the lumped model is applicable (such as for a highly conductive and relatively small body subject to a large external thermal resistance), the assumption of a uniform temperature is justified. Being $E$ the body's internal energy, the First Law of Thermodynamics dictates a macroscopic balance to the body in Fig. 2.21:

$$\dot{Q} = \frac{\mathrm{d}E}{\mathrm{d}\theta} \qquad (2.71)$$

---

Some properties apply: $\text{erf}(0) = 0$; $\text{erf}(0) = 0$; $\text{erf}(\infty) = 1$; $\text{erf}(-x) = -\text{erf}(x)$.
We can approximate $\text{erf}(x)$ [5]:

$$\text{erf}(x) \approx 1 - \frac{1}{(1 + a_1 x + a_2 x^2 + a_3 x^3 + a_4 x^4)^4} \qquad \text{(maximum error: } 5 \times 10^{-4}\text{)}$$

with $a_1 = 0.278393$, $a_2 = 0.230389$, $a_3 = 0.000972$, $a_4 = 0.078108$.

On the other side, as in Eq. (2.32)

$$-\dot{q}_\lambda = \dot{q}_h = \overline{h}\left(T - T_\infty\right), \text{ for } x = -L; \quad \dot{q}_\lambda = \dot{q}_h = \overline{h}\left(T - T_\infty\right), \text{ for } x = L$$
$$(2.32)$$

at its external surface an energy balance gives for an incoming heat flux:

$$\dot{Q} = \overline{h}A\left(T_\infty - T\right) \tag{2.72}$$

The change in the energy budget of the body can be interpreted in terms of the Equation of state for an incompressible solid:

$$\frac{dE}{d\theta} \equiv \rho c_p V \frac{dT}{d\theta} \tag{2.73}$$

Substituting these last two relationships into the First Law of Thermodynamics, one has:

$$-\frac{\overline{h}A}{\rho c_p V}\left(T - T_\infty\right) = \frac{dT}{d\theta} \tag{2.74}$$

which can be manipulated by separating the variable and after change of integration argument (i.e., $dT = d(T - T_\infty)$, since $T_\infty$ is constant):

$$-\frac{\overline{h}A}{\rho c_p V}d\theta = \frac{d\left(T - T_\infty\right)}{\left(T - T_\infty\right)} \tag{2.75}$$

Integration by using the initial condition $T = T_i$, for $\theta = 0$ yields:

$$\ln\frac{T - T_\infty}{T_i - T_\infty} = -\frac{\overline{h}A}{\rho c_p V}\theta \tag{2.76}$$

that is, temperature as function of $\theta$, only:

$$\frac{T(\theta) - T_\infty}{T_i - T_\infty} = \exp\left(-\frac{\overline{h}A}{\rho c_p V}\theta\right) \tag{2.77}$$

As the argument of the exponential must be dimensionless, the group $\rho c V / \overline{h} A$ is called a *characteristic time* or the *time constant* of the unsteady conduction process, driven by internal conduction and external convection; as expected, it would take longer for the body to reach equilibrium with the surrounding fluid when its *thermal capacitance* $\rho c V$ is large, or the *external flux* $\overline{h}A$ (when the geometry is fixed) is small. These two statements amount to large characteristic times, which is opposed to cases with small characteristic times (Fig. 2.25).

**Fig. 2.25** Different temperature progress with different characteristic times

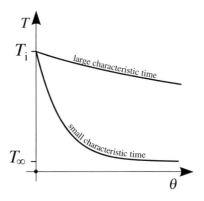

## 2.4 Dimensionless Equation of Conduction

The purpose of dimensional analysis is to identify the controlling dimensionless parameters that are commonly used in the applications of conduction heat transfer. This development describes the **general unsteady regime model** that was anticipated earlier.

### 2.4.1 Dimensional Analysis

For our scopes, it is interesting to focus on the full model of unsteady heat conduction, by means of its governing PDE

$$\frac{\partial^2 T}{\partial x^2} = \frac{1}{\alpha} \frac{\partial T}{\partial \theta} \tag{2.68}$$

and its integrating relationships, with the results conveniently presented in graphical form for rapid engineering calculations. This tool allows to come to a more concise formulation and is commonly employed in various engineering fields.[40]

Let us start by choosing suitable dimensionless independent and dependent variables, relative to the infinite plate (such as the one in Fig. 2.14).

With the nomenclature employed so far, the $BCs$ are:

$$\frac{\partial T(0, \theta)}{\partial x} = 0 \text{ for } \theta > 0; \qquad -\lambda \frac{\partial T(L, \theta)}{\partial x} = \overline{h} \left[ T(L, \theta) - T_\infty \right] \text{ for } \theta > 0 \tag{2.78}$$

---

[40] Another technique can be employed, arriving to the same results, consisting in finding pertinent dimensionless numbers from experimental data, through application of the *Buckingham $\pi$ theorem* [5] (named after American physicist E. BUCKINGHAM, but first proved in the late nineteenth century by French mathematician J. BERTRAND).

while the *initial condition* is:

$$T(x, 0) = T_i \text{ for } 0 \leq x \leq L \tag{2.79}$$

Suitable arbitrary dimensional constants for the desired $T = T(x, \theta)$ are $x_0$, $\theta_0$ and $\Delta T_0$, to obtain:

$$x^* \equiv \frac{x}{x_0}, \quad \theta^* \equiv \frac{\theta}{\theta_0}, \quad T^* \equiv \frac{T - T_\infty}{\Delta T_0} \tag{2.80}$$

The use of the $T - T_\infty$ difference for the dimensionless temperature is suggested by the fact that it is indeed the driving force for the thermal flux. Next, we can proceed constructing the appropriate derivatives with the following four phases:

1. Since it is $\partial x^* = \frac{\partial x}{x_0}$, one has:

$$\frac{\partial T}{\partial x} = \frac{\partial T}{\partial x^*} \frac{\partial x^*}{\partial x} = \frac{\partial T}{\partial x^*} \frac{\partial x}{\partial x} \frac{1}{x_0} = \frac{1}{x_0} \frac{\partial T}{\partial x^*} \tag{2.81}$$

and

$$\frac{\partial^2 T}{\partial x^2} = \frac{\partial}{\partial x} \left( \frac{1}{x_0} \frac{\partial T}{\partial x^*} \right) = \frac{1}{x_0} \frac{\partial}{\partial x^*} \left( \frac{1}{x_0} \frac{\partial T}{\partial x^*} \right) = \frac{1}{x_0^2} \frac{\partial^2 T}{\partial x^{*2}} \tag{2.82}$$

2. Since it is $\partial \theta^* = \frac{\partial \theta}{\theta_0}$, one has:

$$\frac{\partial T}{\partial \theta} = \frac{\partial T}{\partial \theta^*} \frac{\partial \theta^*}{\partial \theta} = \frac{\partial T}{\partial \theta^*} \frac{\partial \theta}{\partial \theta} \frac{1}{\theta_0} = \frac{1}{\theta_0} \frac{\partial T}{\partial \theta^*} \tag{2.83}$$

3. Since it is $\partial T^* = \frac{1}{\Delta T_0} \partial T$ and $\partial^2 T^* = \frac{1}{\Delta T_0} \partial^2 T$, one has:

$$\frac{\partial^2 T^*}{\partial x^{*2}} = \frac{1}{\Delta T_0} \frac{\partial^2 T}{\partial x^{*2}} \tag{2.84}$$

and

$$\frac{\partial T^*}{\partial \theta^*} = \frac{1}{\Delta T_0} \frac{\partial T}{\partial \theta^*} \tag{2.85}$$

4. Let us substitute these findings in Eqs. (2.82) and (2.83), having at the end:

$$\frac{\partial^2 T}{\partial x^2} = \frac{1}{x_0^2} \Delta T_0 \frac{\partial^2 T^*}{\partial x^{*2}} \tag{2.86}$$

and

$$\frac{\partial T}{\partial \theta} = \frac{1}{\theta_0} \Delta T_0 \frac{\partial T^*}{\partial \theta^*} \tag{2.87}$$

So, the governing equation in primitive variables Eq. (2.68) turns to be:

$$\frac{\Delta T_0}{x_0^2}\frac{\partial^2 T^*}{\partial x^{*2}} = \frac{1}{\alpha}\frac{\Delta T_0}{\theta_0}\frac{\partial T^*}{\partial \theta^*} \tag{2.88}$$

and, multiplying both sides by $\frac{x_0^2}{\Delta T_0}$:

$$\frac{\partial^2 T^*}{\partial x^{*2}} = \frac{1}{\alpha}\frac{x_0^2}{\theta_0}\frac{\partial T^*}{\partial \theta^*} \tag{2.89}$$

Moreover, if we choose to simplify further, we can usefully set $x_0 = L$, $\Delta T_0 = T_i - T_\infty$ and $\theta_0 = L^2/\alpha$, coming up with:

$$\boxed{\frac{\partial^2 T^*}{\partial x^{*2}} = \frac{\partial T^*}{\partial \theta^*}} \quad \text{for } 0 \le x^* \le 1 \tag{2.90}$$

$$\frac{\partial T^* (0, \theta^*)}{\partial x^*} = 0 \quad \text{for } \theta^* > 0 \tag{2.91}$$

$$-\frac{\partial T^* (1, \theta^*)}{\partial x^*} = \frac{\overline{h}L}{\lambda}T^*\left(1, \theta^*\right) \quad \text{for } \theta^* > 0 \tag{2.92}$$

$$T^*(x^*, 0) = \frac{T_i - T_\infty}{\Delta T_0} = 1 \quad \text{for } 0 \le x^* \le 1 \tag{2.93}$$

Using the Biot number Bi defined earlier

$$\text{Bi} \equiv \frac{\overline{h}L}{\lambda} \tag{2.46 revisited}$$

and introducing the dimensionless **Fourier number** Fo[41]:

$$\boxed{\text{Fo} \equiv \frac{\theta\alpha}{L^2}} \tag{2.94}$$

the *BCs* of governing Eq. (2.89) become:

$$\frac{\partial T^* (0, \text{Fo})}{\partial x^*} = 0 \quad \text{for } \theta^* > 0 \tag{2.95}$$

---

[41] After the aforementioned scientist J.- B. FOURIER.

and

$$\frac{\partial T^*(1, \text{Fo})}{\partial x^*} = -\text{Bi}\, T^*(1, \text{Fo}) \quad \text{for } \theta^* > 0 \tag{2.96}$$

while the *initial condition* is still:

$$T^*(x^*, 0) = 1 \quad \text{for } 0 \leq x^* \leq 1 \tag{2.97}$$

The analytical (exact) solution of Eq. (2.90), with its boundary and initial conditions Eqs. (2.95–2.97), can be given in a series form. Laying the development of this technique outside the scope of this book, we present the result as follows:

$$T^* = \sum_{n=1}^{\infty} C_n \exp\left(-\zeta_n^2 \text{Fo}\right) \cos\left(-\zeta_n x^*\right) \tag{2.98}$$

with

$$C_n = \frac{4 \sin \zeta_n}{2\zeta_n + \sin(2\zeta_n)} \tag{2.99}$$

and the discrete values of the *eigenvalues* $\zeta_n$ are positive roots of the *transcendental equation*

$$\zeta_n \tan \zeta_n = \text{Bi} \tag{2.100}$$

The first roots, useful to compute a solution, are usually found in tabular form [6].

In lieu of Eq. (2.98), usually a graphical form is preferred in order to help grasp the fundamental ideas concerning the unsteady state of conduction and its characteristic dimensionless numbers. In the end, we see that the function we sought was, **in dimensional form**:

$$T = T\left(x, \theta, L, \alpha, \overline{h}, \lambda, T_\infty, T_i\right) \tag{2.101}$$

while we have obtained, **in dimensionless form**:

$$\boxed{T^* = T^*\left(x^*, \text{Fo}, \text{Bi}\right)} \tag{2.102}$$

with the resulting ease of discussion.

Some algebraic manipulation will clarify on the Fo number meaning:

$$\text{Fo} \equiv \frac{\theta \lambda}{\rho c_p L^2} = \frac{\frac{\lambda}{L} L^2}{\frac{\rho c_p L^3}{\theta}}$$

$$= \frac{\text{conductive power through the surface } L^2}{\text{conductive power accumulated in the volume } L^3} \tag{2.103}$$

Then, the Fo number can be seen as a *dimensionless time*, that is, the ratio between the physical time and a time constant of the process driven by internal conduction.

It is clear that, for greater Fo, a larger penetration of the thermal flux would results, in the time interval under scrutiny.

Fo numbers can be used for other geometries, as well, such as cylinder and sphere [4, 6].

### 2.4.2 A Graphical Tool

The general unsteady regime model, that is, the dimensionless temperature distribution and progress which is solution of Eq. (2.90), for an infinite plate exposed to convection heat transfer by the side, can be calculated in form of $[T(x, \theta) - T_\infty] / [T_i - T_\infty]$ by means of the *Heisler Graphs* [4]. By using the superposition of effects, one has:

$$
\left[\frac{T(x, \theta) - T_\infty}{T_i - T_\infty}\right]_{\text{Plate}} = \left[\frac{T(x, \theta) - T_\infty}{T_0(\theta) - T_\infty}\right]_{\text{Graph } A} \times \left[\frac{T_0(\theta) - T_\infty}{T_i - T_\infty}\right]_{\text{Graph } B} \quad (2.104)
$$

with $T_0$ the temperature at the mid-plane of a plate of thickness $2L$. The two temperature groups on the right-hand side of Eq. (2.104), read on Graphs $A$ (Fig. 2.26) and $B$ (Fig. 2.27), respectively: for the former, in function of the Fourier number Fo and the inverse of the Biot number Bi; for the latter, in function of $1/\text{Bi}$ and the dimensionless abscissa $x/L$.

**Fig. 2.26** Graphs for determining the temperature history of a plate suddenly immersed in a fluid at $T_\infty$. Graph A: relationship between the temperature in any plane $x$ and the temperature at the mid-plane $T_0$ (for $x = 0$). $L$ is the plate mid-thickness

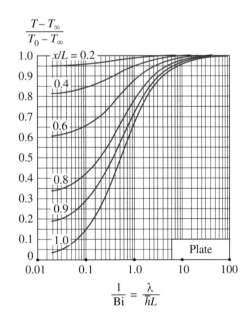

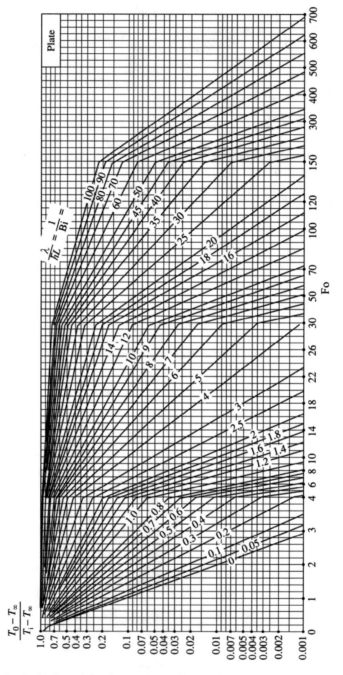

**Fig. 2.27** Graphs for determining the temperature history of a plate suddenly immersed in a fluid at $T_\infty$. Graph $B$: temperature at the mid-plane $T_0$ (for $x = 0$)

One starts with Graph $B$ (Fig. 2.27) by fixing the physical time $\theta$ in which the solution is wanted: the second temperature group on the right-hand side of Eq. (2.104) is computed (knowing the initial temperature $T_i$ and the fluid temperature $T_\infty$) by entering the Fo value and matching with the given conductive/convective mechanism identified by $1/\mathrm{Bi}$. Then the given $1/\mathrm{Bi}$ is entered again in Graph $A$ (Fig. 2.26) and the first temperature group on the right-hand side of Eq. (2.104) is computed, matching with the desired plate depth $x/L$. The curves appear as lines due to the exponential nature of the solution, plotted on a semilogarithmic frame.

Heisler Graphs exist for infinite cylinders and spheres [4, 6].

## 2.5  Numerical Solution of Conduction

We have seen that the driving Equation for conduction, such as the following

$$\rho c_p \frac{\partial T}{\partial \theta} = \lambda \left( \frac{\partial^2 T}{\partial x^2} + \frac{\partial^2 T}{\partial y^2} \right) \pm \dot{e}''' \tag{2.21}$$

or similar ones in other coordinate systems, must be solved to determine the distribution of temperature.

We have outlined briefly in Chap. 1 the differences among the two analysis methods: experiments versus computations. We carried out so far a number of analytical (exact) solutions, presented few others without implementation, and examined the use of graphic tools for simple geometries. But for cases closer to the application world,[42] the use of numerical solutions offered by dedicated software, with the availability of vast and affordable computing power, represents an important solution tool such that the analytical methods seen so far are rarely employed in practice. In other words, the integration of the governing differential equation can be attacked by means of a numerical solution for any complex problems.

In the present Section, the essential ideas underlying the numerical solution of heat conduction problems will be therefore presented. These results may be applied readily to the diffusion of any other scalar, so to anticipate the study that will be completed in Chap. 5. Mathematics is more tractable for heat conduction than for other transport phenomena, the numerical form of the equations usually tending to be well-conditioned and stable unlike with the momentum transfer. This topic therefore is usually the starting point to those wishing to learn the art of virtual processing for any PCB mechanism at stake.

Let us turn then to a method to obtain the numerical (approximate) solution of these Equations, with a special look to the data structure.[43] The method, called a

---

[42]Numerical solution of heat conduction is particularly useful when the shape of the $CV$ is irregular, when its properties vary with space and position, and for nonlinear or non-uniform boundary conditions.

[43]This data structure will be exploited later to study the numerical solution of fluid flow.

*discretization*, consists in a way of transforming the PDEs into a system of *algebraic* equations. The approximations are applied to small domains in space and/or time so the numerical solution would provide results at *discrete locations* in space and time, called *numerical grids*, dividing the physical domain into a finite number of subdomains or *cells* (volumes, elements, etc.) and times. The scope of this Section is to persuade the reader with no specialized background that **numerical methods can be implemented with confidence, for the solution of heat conduction problems**.

## 2.5.1   Discretization

The basic assumption of this kind of numerical method is that the values of temperature are determined at the numerical grid, only. When a suitable grid made of lines drawn parallel to system coordinates is superimposed to the $CV$ (now the calculation domain), these locations are identified by the intersections of the grid, called *grid points*. With the number of grid points increasing greatly (as allowed by the available computing power), the solution of the algebraic equation system (for each item of the chosen grid point set) approaches the exact solution of the corresponding differential equation.

A suitable way to discretize the differential equation is by the **Finite Volume method** (FV). This method, inspired by the work of Patankar [7] among others, will be exercised for every transport mechanism in this book. FV is a brilliant combination of mathematical derivation and physical insight. The calculation domain is divided into a number of non-overlapping smaller finite volumes such that there is one $CV$ surrounding each grid point. The governing PDE for the transport of energy, then, can be integrated over each $CV$, for each point, yielding an algebraic equation containing the value of temperatures for that cluster of points surrounding the subject grid point.

## 2.5.2   Steady-State Conduction, 1-D

### 2.5.2.1   The Grid and the Algebraic Equation for Internal Points

Let us recall the fundamental Equation of conduction in 1-D in the *steady-state* version

$$\frac{d^2 T}{dx^2} + \frac{\dot{e}'''}{\lambda} = 0 \tag{2.31}$$

A generic cluster of grid points surrounding $P$ is presented in Fig. 2.28. Integrating Eq. (2.31) over the $CV$ of $P$, we get[44]

---

[44]For a non-uniform medium, $\lambda$ should be evaluated at the faces $w$ and $e$; if $\dot{e}'''$ is no longer uniform, it should be subject to integration across the $CV$. Non-uniformity can be healed by proper gridding increase across the interface.

**Fig. 2.28** A cluster of grid points ($W$ and $E$) surrounding the subject grid point, $P$, for the 1-D problem. The hollow circles mark the interface points ($w$ and $e$), and the dashed lines mark the $CS$ at those points

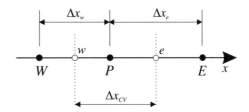

$$\lambda \left. \frac{dT}{dx} \right|_e - \lambda \left. \frac{dT}{dx} \right|_w + \dot{e}''' \int_w^e dx = 0 \qquad (2.105)$$

To evaluate the derivatives, we will employ no formal way such as in Sect. 2.3.3 (p. 31); we will exploit instead a *finite difference* representation[45] for the first derivative of $T$ at $x = x_0$:

$$\left. \frac{dT}{dx} \right|_{x_0} \approx \frac{T(x_0 + \Delta x) - T(x_0)}{\Delta x} \qquad (2.106)$$

so that Eq. (2.105) can be rewritten, for sufficiently small $\Delta x$'s:

$$\boxed{\lambda \frac{T_E - T_P}{\Delta x_e} - \lambda \frac{T_P - T_W}{\Delta x_w} + \dot{e}''' \Delta x_{CV} = 0} \qquad (2.107)$$

It is useful to cast Eq. (2.107) in the following form:

$$\boxed{a_P T_P + a_E T_E + a_W T_W = b} \qquad (2.108)$$

where

$$a_E = \frac{\lambda}{\Delta x_e} \qquad (2.109a)$$

$$a_W = \frac{\lambda}{\Delta x_w} \qquad (2.109b)$$

---

[45]Let us recall the definition of the derivative of $T(x)$ at $x = x_0$: $\frac{dT}{dx} \equiv \lim_{\Delta x \to 0} \frac{T(x_0 + \Delta x) - T(x_0)}{\Delta x}$. Here, if $T$ is continuous, it is expected that $\frac{T(x_0 + \Delta x) - T(x_0)}{\Delta x}$ (also called a **forward difference scheme**) will be a "reasonable" approximation to $\frac{dT}{dx}$ for a "sufficiently" small but finite $\Delta x$. Indeed, just like in Eq. (2.22) but across a finite $\Delta x$ interval this time, the Taylor'sa series expansion is written as $T(x_0 + \Delta x) = T(x_0) + \frac{dT(x_0)}{dx} \Delta x +$ a truncation error [3].

$$a_P = -(a_E + a_W) \tag{2.109c}$$

$$b = -\dot{e}''' \Delta x_{CV} \tag{2.109d}$$

In case of anisotropic thermal conductivity $k$, Eqs. (2.109a) and (2.109b) modify into:

$$a_E = \frac{\lambda_e}{\Delta x_e} \tag{2.110a}$$

$$a_W = \frac{\lambda_w}{\Delta x_w} \tag{2.110b}$$

While forming the final algebraic system, a general form like Eq. (2.108) can be cast for any *internal grid point*, that is, for any situation depicted in Fig. 2.28.

### 2.5.2.2   The Equation for Boundary Points

Then if we consider the sample, complete geometry reported in Fig. 2.29, we have to write two more Equations, involving the *boundary grid points* temperatures or their derivatives. Indeed, as discussed already with Eq. (2.22), on a given boundary an appropriate BC must be specified. For example, if the value of the boundary temperature in $B$, $T_B$, is known (Fig. 2.29), a standard equation can be cast as usual, but if $T_B$ is not given, we need to integrate Eq. (2.31) over the $CV$ of boundary grid point $B$. Let $\dot{q}_B$ be the thermal flux applied at the left face of $CV$ of $B$. We get for this grid point, instead of Eq. (2.107):

$$\dot{q}_B - \lambda \frac{T_B - T_I}{\Delta x_i} + \dot{e}''' \Delta x_{CV} = 0 \tag{2.111}$$

The required equation for $T_B$ becomes

$$a_B T_B + a_I T_I = b \tag{2.112}$$

where

$$a_I = \frac{\lambda}{\Delta x_i} \tag{2.113a}$$

$$a_B = -a_I \tag{2.113b}$$

$$b = -\dot{q}_B - \dot{e}''' \Delta x_{CV} \tag{2.113c}$$

Here, with reference to the general Eq. (2.108), $a_I$ and $a_B$ replace $a_E$ and $a_P$, respectively. In case of variable $\lambda$, Eq. (2.113a) modifies into:

$$a_I = \frac{\lambda_i}{\Delta x_i} \tag{2.114}$$

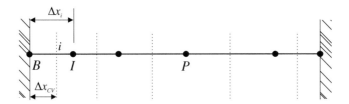

**Fig. 2.29**  A complete cluster of grid points for a conduction 1-D problem. Grid point $I$ is an internal point, adjacent to boundary point $B$. The dashed lines mark the $CS$ of each $CV$

A similar development will result in case of specification of the boundary thermal flux via a given convective transmittance at point $B$, $h$.

### 2.5.2.3   Structure and Solution of the Algebraic Equation System

After gathering the $n$ solving equations relative to the grid points, a *matrix representation* can be provided:

$$
\begin{bmatrix}
a_{P1} & a_{E1} & 0 & 0 & 0 & 0 \\
a_{W2} & a_{P2} & a_{E2} & 0 & 0 & 0 \\
0 & a_{W3} & a_{P3} & a_{E3} & 0 & 0 \\
\hdotsfor{6} \\
0 & 0 & 0 & a_{W(n-1)} & a_{P(n-1)} & a_{E(n-1)} \\
0 & 0 & 0 & 0 & a_{Wn} & a_{Pn}
\end{bmatrix}
\begin{bmatrix}
T_1 \\ T_2 \\ T_3 \\ \cdots \\ T_{n-1} \\ T_n
\end{bmatrix}
=
\begin{bmatrix}
b_1 \\ b_2 \\ b_3 \\ \cdots \\ b_{n-1} \\ b_n
\end{bmatrix}
\qquad (2.115)
$$

We see that all the nonzero coefficients $a$ align themselves along three diagonals of the coefficient matrix.[46] One convenient procedure to solve such system, in this case, is the *Thomas algorithm*[47] [3] involving the following sequence:

1. the system is first put into an upper triangular form, by computing the new $a_{Pi}$ by

$$
a_{Pi} = a_{Pi} - \frac{a_{Wi}}{a_{P(i-1)}} a_{E(i-1)} \qquad i = 2, 3, \cdots, n \qquad (2.116)
$$

and the new $b_i$ by

$$
b_i = b_i - \frac{a_{Wi}}{a_{P(i-1)}} b_{i-1} \qquad i = 2, 3, \cdots, n \qquad (2.117)
$$

---

[46]It is clear, from inspection of coefficient equations like Eq. (2.108), that the matrix is diagonally dominant (also called the Scarborough criterion, as proposed by the American J.B. SCARBOROUGH, at mid-twentieth century), with coefficients that must be always discordant with those of the lateral nonzero diagonals.

[47]After English physicist and mathematician L.H. THOMAS in mid-twentieth century.

2. the unknowns $T$'s are computed from back-substitution, using first $T_n = b_n/a_{Pn}$
   for the last row and then

$$T_k = \frac{b_k - a_{Ek}T_{k+1}}{a_{Pk}} \qquad k = n-1, n-2, \cdots, 1 \qquad (2.118)$$

for the remaining rows.

We may use this algorithm later, and in every circumstance, our solving system appears with this data structure.


#### 2.5.2.4   Treatment of Nonlinearities

So far we consider Eq. (2.31) along with its simplifying assumptions of material properties and source/sink term independent on temperature. When this simplification cannot apply (as in common situations), the coefficients of Eq. (2.115) will depend on the variable itself and we end up with a *nonlinear equation system* instead. In this case, we can handle the situation by **iteration**, involving the following operation cycle for the variable $T$:

1. start with a guess or estimate $T^*$ distribution at all grid points;
2. from these $T^*$, calculate tentative values of the coefficients in the algebraic
   system Eq. (2.115);
3. solve the nominally linear system to get new values of $T$;
4. with these $T$ as better $T^*$ guesses, return to Step 2 and repeat the process until
   further iterations cease to produce any significant changes in the distribution of
   $T$.

This final unchanging state is the *convergence of the iterations* and is the solution of the nonlinear system. It is, however, possible that successive iterations would not ever converge to a solution. The variable values may steadily drift or oscillate with increasing amplitude. This process, at the opposite of convergence, is called *divergence*.

Care should also be exercised to ensure that, in case of a strong source/sink term (that would substantially modify the $b_i$ coefficients in the right-hand side column of Eq. (2.115)), the rule recalled by Note 46 (diagonal dominance) is always enforced. In such a case, it is desirable to control the rate at which variables are changing during iterations. Moreover, if the initial guess is far from the solution, we may get large oscillations in the computed variable during the first few iterations, making it difficult for the iteration to proceed.

The simplest method to favor convergence is by an *under-relaxation* procedure. With this, at the Step 2 above, $T_P$ is computed by using Eq. (2.108):

$$T_P = \frac{b - a_E T_E - a_W T_W}{a_P} \qquad (2.119)$$

The change of $T_P$ that we want to monitor is given by $\left( \frac{b - a_E T_E - a_W T_W}{a_P} - T_P^* \right)$, and we wish to make $T_P$ change only by a fraction $\gamma$ (the under-relaxation coefficient) of this:

$$T_P = T_P^* + \gamma \left( \frac{b - a_E T_E - a_W T_W}{a_P} - T_P^* \right) \quad \text{with} \ \ 0 \leq \gamma \leq 1 \qquad (2.120)$$

Thus, Eq. (2.108) can be replaced by

$$\frac{a_P}{\gamma} T_P + a_E T_E + a_W T_W = b + \frac{1 - \gamma}{\gamma} a_P T_P^* \qquad (2.121)$$

With this approach we ensure that, upon convergence $T_P = T_P^*$, the standard Eq. (2.108) is correctly applied. Also, with Eq. (2.121) we are again enforcing the diagonal dominance requirement.

The value of $\gamma$ which reduces the computational load of the iterations while ensuring convergence depends strongly on the nature of the solving system (driving equations and related BCs for the given PCB problem), on how strong the nonlinearities are, on the grid strategy (an increased gridding where variable gradients are larger would reduce their rate of change), and so on. A value close to unity allows the solution to move quickly toward convergence, but may be more prone to divergence. A low value keeps the solution close to the initial guess, but keeps the solution from diverging. Intuition and experience are employed in choosing the appropriate $\gamma$ value for each case at stake.

### 2.5.3  Unsteady-State Conduction

Now, let us turn the unsteady-state version of the fundamental Equation of conduction in 1-D and *constant properties*

$$\frac{\partial^2 T}{\partial x^2} = \frac{1}{\alpha} \frac{\partial T}{\partial \theta} \qquad (2.68)$$

This time, we need to obtain a solution by "Marching in time," that is, by proceeding along a series of "time steps" or *time discretization*, from a given distribution of temperature to the final one. In other words, in a typical time step, starting from the grid point values of $T^0$ at time $\theta$, we will evaluate the "new" values $T$ at the time $\theta + \Delta\theta$.

Integrating Eq. (2.68) over the $CV$ of $P$ (Fig. 2.28), we get

$$\lambda \int_\theta^{\theta+\Delta\theta} \left( \frac{\partial T}{\partial x}\bigg|_e - \frac{\partial T}{\partial x}\bigg|_w \right) d\theta = \rho c_p \int_w^e \int_\theta^{\theta+\Delta\theta} \frac{\partial T}{\partial \theta} dx d\theta \qquad (2.122)$$

Here, the evaluation of derivatives of terms on the left-hand side along $x$ has been already carried out, according to Eqs. (2.105–2.107). Now, as a supplement of the FV characteristics declared in Sect. 2.5.1 (p. 56), we will assume that the $T_P$ value prevails throughout the subject $CV$. Then, with Eq. (2.106) in mind:

$$\rho c_p \int_w^e \int_\theta^{\theta+\Delta\theta} \frac{\partial T}{\partial\theta} d\theta dx = \rho c_p \Delta x_{CV}(T_P - T_P^0) \qquad (2.123)$$

and Eq. (2.122) becomes

$$\boxed{\lambda \int_\theta^{\theta+\Delta\theta} \left( \frac{T_E - T_P}{\Delta x_e} - \frac{T_P - T_W}{\Delta x_w} \right) d\theta = \rho c_p \Delta x_{CV}(T_P - T_P^0)} \qquad (2.124)$$

Nothing has been said so far on the time step at which the temperature on the left-hand side of Eq. (2.115) should be computed. If we assume that the task of advancing in time is performed solely by the term on the right-hand side, and the heat exchange between grid point is entirely computed by means of the new temperatures at the left-hand side, it is then

$$\int_\theta^{\theta+\Delta\theta} T d\theta = T \Delta\theta \qquad (2.125)$$

In this way, all temperatures (except $T_P^0$) are assumed at the new time $\theta + \Delta\theta$; the discrete form Eq. (2.124) is called the *fully implicit scheme*, and we come up with

$$\boxed{a_P T_P + a_E T_E + a_W T_W = b} \qquad (2.126)$$

where

$$a_E = \frac{\lambda}{\Delta x_e} \qquad (2.127a)$$

$$a_W = \frac{\lambda}{\Delta x_w} \qquad (2.127b)$$

$$a_P^0 = -\frac{\rho c_p \Delta x_{CV}}{\Delta\theta} \qquad (2.127c)$$

$$a_P = -\left( a_E + a_W + a_P^0 \right) \qquad (2.127d)$$

$$b = a_P^0 T_P^0 \qquad (2.127e)$$

In case of variable properties, the coefficients are modified as specified earlier. Aspects of the procedure adopted for the steady-state conduction, such that boundary conditions and system solution, are applicable to the unsteady situation. As for the temperature-dependent conductivity and source/sink term case, the iterative procedure seen above also applies, by using the new (in the sense of "time") values of $T_P$.

### 2.5.4 Extension to 2-D Geometry and Solver Types

The ideas behind the development of Eq. (2.126) can be readily extended to rectangular 2-D and 3-D geometries, as well as for other geometries, with no further difficulty. For example, with reference to the 2-D rectangular geometry in Fig. 2.30, the fundamental Equation of conduction Eq. (2.21) can be considered again, and its discrete equation turns to be

$$a_P T_P + a_E T_E + a_W T_W + a_N T_N + a_S T_S = b \qquad (2.128)$$

where

$$a_E = \frac{\Delta y_{CV}}{\Delta x_e} \qquad (2.129a)$$

$$a_W = \frac{\Delta y_{CV}}{\Delta x_w} \qquad (2.129b)$$

$$a_N = \frac{\Delta x_{CV}}{\Delta x_n} \qquad (2.129c)$$

$$a_S = \frac{\Delta x_{CV}}{\Delta x_s} \qquad (2.129d)$$

$$a_P^0 = -\frac{\rho c_p \Delta x_{CV} \Delta y_{CV}}{\lambda \Delta \theta} = -\frac{1}{Fo} \qquad (2.129e)$$

$$a_P = -(a_E + a_W + a_N + a_S + a_P^0) \qquad (2.129f)$$

$$b = a_P^0 T_P^0 - \frac{\dot{e}'''}{\lambda} \Delta x_{CV} \Delta y_{CV} \qquad (2.129g)$$

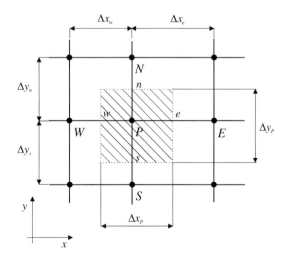

**Fig. 2.30** A generic cluster of grid points ($W$, $E$, $N$, and $S$) surrounding the subject grid point $P$, for the 2-D problem. The dashed lines mark the $CS$ (faces at $w$, $e$, $n$, and $s$). $CV$ lengths are indicated

with Fo a *grid Fourier number*, when λ is constant.[48] The related BCs are also straightforward.

The coefficient matrix resembles Eq. (2.115), with nonzero values aligned along 5 diagonals.[49] In order to solve it, a combination of the concepts like the ones presented for the 1-D discrete equation is possible: a variety of fast *algebraic solvers* exist for the user to pick and exploit. Generally, *direct solvers* (those requiring no iteration) or *iterative solvers* are available by means of numerical analysis software. The adoption of the solver type is usually dictated by a trade-off between *performance* (computational time required to yield the viable solution) and *robustness* (ability to come to a solution even in presence of strong nonlinearities). Problems with modest grid sizes usually enjoy from direct solver implementation. Direct solvers can be very robust but their memory requirement scales considerably with the number of the unknowns. For large 4-D (unsteady) problems, efficient iterative solvers exist instead, with a wide variety of solver options to reduce the computational time required.

We will soon realize that, as we dig in more transport phenomena mechanisms, the choice and the strategy to adopt for the numerical solution become more stringent.

## 2.5.5  Grid Types

We have seen that the starting point, to set up a numerical solution of a conduction heat transfer problem, is the discretization of the physical domain. Very often the numerical grid or *mesh* is devised in more elaborate form than the ones implied so far. When the domain is *regular in shape* (rectangles, cubes, cylinders, spheres), these can be meshed by regular grids, as shown in Fig. 2.31, ① and ②. The grid lines are orthogonal to each other and conform to the boundaries of the domain: these meshes are also sometimes called *orthogonal grid*.

For many practical problems, however, the domain of interest is irregularly shaped and regular meshes may be improper. An example is shown in Fig. 2.31, ③: here, grid lines are not necessarily orthogonal to each other and curve to conform to the irregular geometry. If regular grids are used in these geometries, *stair-stepping* occurs at domain boundaries, as shown in Fig. 2.32.

The meshes shown in Figs. 2.31 and 2.32 are examples of *structured grid*, which are usually employed in numerical solution of conduction. Being the simplest arrangement, this grid consists in families of grid lines with the property that members of a single family do not cross each other and cross each member of the other families only once. This allows the lines of a grid to be numbered consecutively and

---

[48]This number should not be confused with the one introduced earlier with Eq. (2.93). The grid Fo is used in computational stability criteria when schemes other than the fully implicit one are adopted.

[49]For 3-D geometries, the coefficient matrix Eq. (2.115) will have nonzero values aligned along 7 diagonals, to include the contributions of the $CV$s along the $z$ coordinate.

**Fig. 2.31** Orthogonal grids
① and ②, and an irregular
grid ③

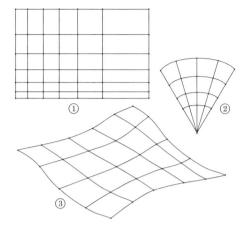

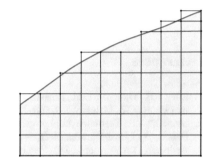

**Fig. 2.32** A case of
stair-stepping of an irregular
boundary meshed with a
regular grid

the position of any grid point to be uniquely identified by a set of two (in 2-D) or three (in 3-D) indices.

The *gridding strategy* should take into account the expected increase of spatial and/or temporal temperature gradients in given zones of the $CV$, by placing more grid points therein. Recalling Note 45, this practice is motivated by the fact that the approximation errors due to discretization need to be minimized. As an example, in Fig. 2.31 ① a local increase of grid points in the lower left corner of the $CV$ is arranged, where a spatial gradient is expected.

## 2.6 Further Reading

- Analytical methods in steady and unsteady conduction. Arpaci, V.S.: Conduction Heat Transfer. Addison-Wesley, Nw York (1966)
- Use of shape factors as a generalized method to develop governing equations in several coordinate systems; Integration of differential equation for the unsteady state; Heisler and Grober charts. Yovanovich, M.M.: Conduction and Thermal

Contact Resistances (Conductances). Pages 3.1–3.33. In: Rohsenow, W.M., Hart-nett, J.P., Cho, Y.I. (Eds.) Handbook of Heat Transfer. McGraw-Hill, New York (1998)

- Issues in modeling; Principles for mathematical modeling; Dimensional analysis. Himmelblau, D.M.: Mathematical Modeling. In: Bisio, A., Kabel, R.L. (Eds.) Scaleup of Chemical Processes. John Wiley & Sons, New York (1985)
- Dimensionless numbers in conduction heat transfer and their meaning. Kuneš, J.: Dimensionless Physical Quantities in Science and Engineering. Elsevier, London (2012)
- Numerical methods for conduction heat transfer. Jaluria, Y., Torrance, K. E.: Computational Heat Transfer. Hemisphere Publishing, Washington (1986)
- Separation of variables; Superposition; Development of finite difference and finite element computations. Myers, G.E.: Analytical Methods in Conduction Heat Transfer. Chapters 3, 4, 8, 9. McGraw-Hill, New York (1971)

# References

1. Liley, P.E., Thomson, G.H., Friend, D.G., Daubert, T.E., Buck, E.: Physical and chemical data. In: Green, D.W., Maloney, J.O. (Eds.) Perry's Chemical Engineers' Handbook, Chap. 2 (1997)
2. Irvine Jr., T.F.: Thermophysical properties. In: Rohsenow, W.M., Hartnett, J.P., Cho, Y.I. (eds.) Handbook of Heat Transfer. McGraw-Hill, New York (1998)
3. Anderson, D.A., Tannehill, J.C., Pletcher, R.H.: Computational Fluid Dynamics and Heat Transfer. Hemisphere Publishing, Washington (1984)
4. Bejan, A.: Heat Transfer. Wiley, New York (1993)
5. Mills, A.F.: Heat Transfer. Richard D. Irwin, Boston (1992)
6. Bergman, T.L., Incropera, F.P., Lavine, A.: Fundamentals of Heat and Mass Transfer. Wiley, New York (2011)
7. Patankar, S.V.: Numerical Heat Transfer and Fluid Flow. Hemisphere Publishing, Washington (1980)

# Chapter 3
# Momentum Transfer

**Abstract** *Transfer of momentum* is the second physical mechanism we encounter, through which the *mechanics of fluids* can be examined, including their interaction with the solid structures that may contain, and limit, them. As the subject implies the new concept of motion of matter, some additional care is needed in the analysis. After a brief reference to the basic physical mechanism, we see that fluid flow occurs when a given *pressure and/or velocity difference* is applied to the fluid. We work out the *governing differential equations* in two main cases, as both microscopic and macroscopic balances are developed leading to the **distribution of the flow velocity vector**. With this mechanism, the subject medium participates directly by taking shape and being deformed due to the inherent phase (fluid, solid) interaction and by being traveled across by heat. Then, the fundamental equations of fluid mechanics are derived, which are pervasive in situations where a detailed description of flow field is needed. The concept of *boundary layer* is then laid out. Finally, the *numerical solution* is continued by exploiting the problem structure that was initiated in the preceding chapter.

## 3.1 Fluid Dynamics: The Underlying Physics and Basic Definitions

### 3.1.1 Fluid Motion and Flow Regimes

Unlike heat, which is related to a more direct definition stemming from the Thermodynamics, *momentum* deserves that we recall its definition from mechanics. The momentum of a fluid is the **product of its mass by its velocity, $m\mathbf{w}$**; as in any dynamic phenomenon, fluid motion is based on Newton's second Law of motion, which states that the vector sum of the external forces $\mathbf{F}$ on an object is equal to the time derivative of its mass multiplied by its velocity:

---

The original version of this chapter was revised: Belated corrections have been incorporated. The correction to this chapter is available at https://doi.org/10.1007/978-3-319-66822-2_7

© Springer International Publishing AG 2018

G. Ruocco, *Introduction to Transport Phenomena Modeling*,

https://doi.org/10.1007/978-3-319-66822-2_3

$$\sum \mathbf{F} = \frac{d(m\mathbf{w})}{d\theta} \qquad (3.1)$$

The *viscous nature of fluids* governs the momentum transfer generating *friction,* within the fluid itself (*fluid friction*) and between a solid surface and its "wetting" fluid (*skin friction*), in both cases when relative motion occurs. The notion of flow deformation due to the perturbing action by the flow's confining surfaces, called the *velocity boundary layer,* is also at stake.

From the point of view of the macroscopic mean or bulk behavior of the fluid, it is customary to subdivide the flows based on their velocity regime, which can be *laminar, turbulent,* or *transitional* (when passing from one regime to the other). Laminar flow occurs when a fluid flows in parallel layers, with no disruption between the layers.[1] At low velocities, the fluid tends to flow without lateral mixing: adjacent layers slide past undisturbed by one another. When a colored dye is injected into the fluid at some point in a tube flow, it follows a well-defined path without appreciable mixing. In other words, there are no cross-currents perpendicular to the direction of flow, nor eddies or swirls of fluids.

Above a certain velocity magnitude threshold, turbulent flow occurs. Turbulence or turbulent flow is a flow regime characterized by chaotic property changes, such as pressure and velocity, both in space and time. The colored dye, in this case, will be distributed over a wide area a short distance downstream from the point of injection. The mixing mechanism consists of rapidly fluctuating vortices or eddies on many dimensional scales that transport fluid particles in an irregular manner and interact with each other. Drag due to boundary layer skin friction increases. The structure and location of boundary layer separation often change, as will be reported later, sometimes resulting in a reduction of overall drag. Flow mixing is greatly enhanced, and the associated mechanisms of heat and mass transfer will be many times more effective than in laminar flow.

The fluid properties are paramount in assessing the flow behavior or regime. To this end, let us anticipate a discriminating parameter. We denote with $\rho_f$ and $\mu$ the fluid *density* and *dynamic viscosity,* respectively; then, we let $L$ a *characteristic length*. We also need a *reference value of* $\mathbf{w}$, in this case the *bulk value of the axial velocity component* $\langle w \rangle$. With these parameters, the flow regime can be assessed by evaluating the dimensionless **Reynolds number**[2]

$$\text{Re} \equiv \frac{\rho_f \langle w \rangle L}{\mu} \qquad (3.2)$$

This number can be interpreted as the *ratio between inertia and viscous forces*. It is impossible to disregard the importance of this dimensionless number, characterizing the hydrodynamic conditions for viscous fluid flow.

---

[1] Also called the flow "blades" or *laminae*.

[2] As proposed by English engineer and physicist O. REYNOLDS, in the late nineteenth century.

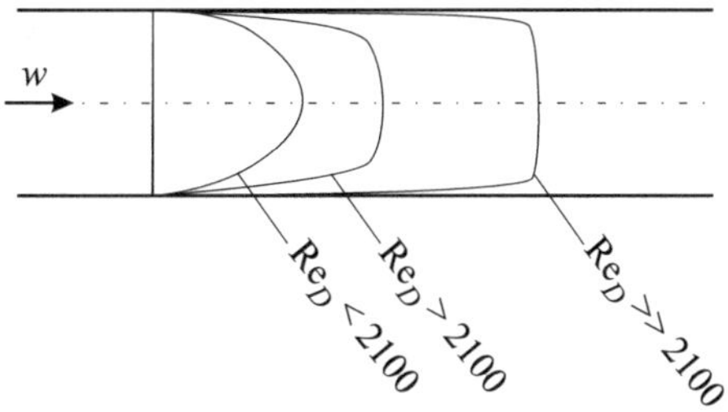

**Fig. 3.1** Sample distributions of $w$ (the axial component of **w**) along the radius of a tube, with varying Reynolds number $\mathrm{Re}_D$

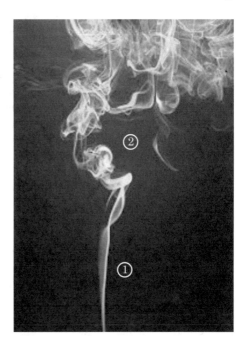

$$w \longrightarrow$$

$\mathrm{Re}_D < 2100$ \qquad $\mathrm{Re}_D > 2100$ \qquad $\mathrm{Re}_D \gg 2100$

**Fig. 3.2** A comparison between laminar ① and turbulent flow ② regimes, for a column of cigarette smoke

For example, measurements on the velocity profile of fluid across a tubing, depending on the flow regime, would be represented as in Fig. 3.1. In this case, Re is based on tube diameter $D$ and would be denoted with $\mathrm{Re}_D$. The profile will be exactly *parabolic* in the laminar regime,[3] while it will grow closer to a generally flatter profile[4] in the central part of the flow, approaching the turbulent regime (in this case, we have assumed that the conventional value of 2100 is the discriminant value for the flow regimes). It is soon evident from Fig. 3.1 the effect of fluid flow deformation, which is the boundary layer itself, especially in the vicinity of the surface. The variety of velocity profiles justifies the need to take a close look to the flow field, to infer on the actual flow process.

---

[3] This will be verified later by studying the related analytical solution.

[4] Also called a *slug flow*.

Another comparison between the two regimes can be inspected with Fig. 3.2, by visualizing the smoke flow rising from a cigarette, in an unconfined geometry. The combustion and release of hot, opaque compounds form a *column of smoke* ①, warmer than the surrounding air. With no external perturbation, the smoke column develops upward, forming an orderly, deterministic laminar flow. At some point, the smoke column cools down and cannot maintain its stability (as the drive offered by free convection weakens) in the upward motion; therefore, it diffuses irregularly in the surrounds with characteristic turbulent *eddies* ②: a turbulent flow for which the details of the motion are never identically repeated.

In the various applications, even under the same geometries and operation conditions, fluids behave differently and a variety of flow situations may occur. To understand this, we will deal again with the inherent microscopic or molecular mechanism, reviewing the concept of *viscosity*, which is the fluid constitutive property, in relation to the momentum transfer. This will lead us to the knowledge of the velocity distribution, paramount in determining the behavior of flow.

Let us proceed with our analysis by taking a close look at the medium or fluid properties that govern the workings of momentum transfer. Specifically, we will focus on the inherent microscopic or molecular mechanism.

### 3.1.2  Newton's Law of Viscosity

#### 3.1.2.1  Velocity Distribution in a Viscous Fluid

Consider a fluid $CV$ flowing between two parallel plates (Fig. 3.3). The flow is defined by a finite depth $L$ along $y$, but it is very wide instead along the other two coordinates. This large width (area), in any $x - z$ plane, will have a size $A$. The fluid and the plates are initially in thermal equilibrium and at rest: the thick line in Fig. 3.3a is parallel to $y$. At a time $\theta = 0$ (Fig. 3.3b), the lower plate is set and maintained at a given constant velocity $W$, by means of a tangential force $F$ parallel to the bottom plate.

A velocity profile $u(y, \theta)$ is therefore created as in Fig. 3.3c. Let us assume that *the flow regime is laminar*: this is determined by the sole force $F$ applied, and no pressure gradients are observed in the flow direction.[5] After some time (Fig. 3.3d), when the steady state is restored,[6] the medium's molecules that touch the bottom plate are found at the same velocity $W$, while those in contact with the top plate remain at rest:

$$u(0, \infty) = W, \quad u(L, \infty) = 0 \tag{3.3}$$

---

[5]This flow is also called a *Couette flow*, after French physicist M.M.A. COUETTE, in the early twentieth century.

[6]For many fluids of technical interest, as will be seen later, after a sufficiently long time, this profile is a linear one.

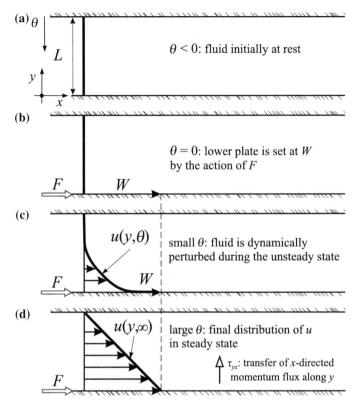

**Fig. 3.3** Variation, during time progress $\theta$ (from top **a**, to bottom **d**), of the velocity profile $u$ in the fluid, due to the dynamic perturbation given by the lower plate set and maintained in motion by force $F$. The direction of the moment flux is evidenced at lower right

### 3.1.2.2 Momentum Flux

Let us denote with $\tau_{yx}$ the **tangential flux**, also called the **shear stress** or shear flux. This is the *pressure* corresponding to the force $F$ applied to the lower plate of extension $A$, in order to set and keep the plate in motion at the velocity $W$. $\tau_{yx}$[7,8] is the *flux of momentum transfer*. It is:

$$\frac{F}{A} = \tau_{yx} \propto \frac{W}{L} \tag{3.4}$$

In the steady state, the force $F$ balances out the dissipative effects inherent to the fluid in motion, as a **net molecular transport of momentum through adjacent fluid sheets** occurs, *in the direction of velocity decrease*. Equation (3.3) evidences

---

[7]Meaning of subscripts: $\tau_{yx}$ is the component of the stress $\tau$ exerting toward $y$ but parallel to $x$.

[8]Shear stress $\tau_{yx}$ must not be mistaken with the *shear rate* or *deformation*, which is $\dot{\gamma} \equiv \partial u / \partial y$.

the existing link between the *momentum flux transferred through the fluid and the observable velocity field*. In other words, the **velocity gradient is the driving force of momentum transfer**. It is $[\tau_{yx}]=N/m^2=Pa$.

### 3.1.2.3  Link Between Stress and Strain

Once a $CS$ and a $CV$ are assigned, the essential problems in studying momentum transfer are

- the *determination of the velocity distribution* in the $CV$
- frequently, the *effect of interaction between phases* at their $CS$ interface. As an example, the fluid resistance (force through a $CS$) during relative motion between two different fluids, or a fluid and a solid

The proportionality parameter inherent to Eq. (3.4) is characteristic of the fluid, linking *cause* (stress) and *effect* (strain, or deformation). When the relationship Eq. (3.4) is a *linear* one (recall Fig. 3.3d and Note 6), then the fluid is called *Newtonian* and the proportionality parameter is a constant called the fluid **dynamic viscosity** $\mu$. Having noted that the flux of momentum transfer is positive when the velocity gradient is negative, we can introduce the differential form of Eq. (3.4) serving to our purposes:

$$\tau_{yx} = -\mu\frac{\partial u}{\partial y} \tag{3.5}$$

This is the **Newton's Law of viscosity**: *the greater the fluid viscosity, the stronger the tangential force necessary to keep the lower plate at the desired velocity.*[9] It is $[\mu]=kg/ms$ (or $Ns/m^2$) in SI units, while a frequent derived unit is the poise P (with 1 P=0.1 Pas).

The ratio of dynamic viscosity over fluid density is defined as the **kinematic viscosity** $\nu$ that measures the *propagation of the dynamic perturbation*:

$$\nu \equiv \frac{\mu}{\rho_f} \tag{3.6}$$

It is $[\nu]=m^2/s$ in SI units,[10] while a frequent derived unit is the stoke St (1 St=1 $cm^2/s$).

---

[9]As proposed by English physicist, mathematician, astronomer, and philosopher I. NEWTON, in the early eighteenth century.

[10]It is worth to note that thermal diffusivity $\alpha$ Eq. (2.9) and kinematic diffusivity $\nu$ share the same units.

### 3.1.3 Driving Material Properties

The material property parameters for this transport mechanism, or *rheological properties* of fluids [1, 2], are the viscosities defined earlier with Eqs. (3.5) and (3.6). For water and air, which are common in the applications and their behavior and, therefore, can be taken as a reference, it is instructive to examine how they vary with temperature: this problem will be at stake later when studying convection heat transfer. For Newtonian fluids such as those, $\mu$ is a property of state: for liquids, it is practically independent on pressure and soon decreases with temperature; for gases, it is approximatively independent on pressure but increases with temperature instead, as in Table 3.1. As for the kinematic viscosity, we see in Table 3.2 that the dependence on temperature is reversed.

As evidenced earlier, the deformable nature of fluids gives rise to stress and strain. The model of Newtonian fluid, presenting a single value of viscosity for a specific temperature but unchanging with the strain rate, represents a good approximation for many liquids and in general for gases. Nevertheless, other situations exist, in which viscosity decreases under mechanical agitation (shear thinning), or just do the opposite (shear thickening). For these fluids, generally called *non-Newtonian*, the linear relationship Eq. (3.4) needs to be properly modified with other models. Usually, the non-Newtonian viscosity is indicated with the specific notation of $\eta$ which is function, with Note 8 in mind, of the shear rate $\dot{\gamma}$. Figure 3.4 summarizes on some possible cases:

1. inviscid fluids ($\mu = 0$)
2. Newtonian fluids
3. pseudoplastic fluids (shear thinning)
4. dilatant fluids (shear thickening)
5. plastic fluids

**Table 3.1** Dynamic viscosity $\mu$ (Pas) for water and air as a function of temperature $T$ (°C)

| $T$ (°C) | Water, sat.liq. | Water, super.steam | Air |
| --- | --- | --- | --- |
| 20 | $993 \times 10^{-6}$ | – | $18.2 \times 10^{-6}$ |
| 100 | $278 \times 10^{-6}$ | $12.4 \times 10^{-6}$ | $21.8 \times 10^{-6}$ |
| 150 | $208 \times 10^{-6}$ | $14.3 \times 10^{-6}$ | $23.9 \times 10^{-6}$ |

**Table 3.2** Kinematic viscosity $\nu$ (m²/s) for water and air as a function of temperature $T$ (°C)

| $T$ (°C) | Water, sat.liq. | Water, super.steam | Air |
| --- | --- | --- | --- |
| 100 | $0.2940 \times 10^{-6}$ | $21.60 \times 10^{-6}$ | $23.05 \times 10^{-6}$ |
| 150 | $0.2270 \times 10^{-6}$ | $27.37 \times 10^{-6}$ | $28.62 \times 10^{-6}$ |

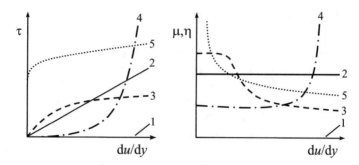

**Fig. 3.4** Sample stress–strain (left) and viscosity–strain (right) relationships, for a variety of situations: 1. inviscid fluids, 2. Newtonian fluids, 3. pseudoplastic fluids, 4. dilatant fluids, 5. plastic fluids

## 3.2  Elementary Viscous Flow: Driving Factors

A great many flow situations deal with the presence of containing or bounding surfaces. From Thermodynamics, it is known that it is the occurrence of viscosity that brings forth the existence of irreversibility due to friction, found between adjacent fluid sheets (*fluid friction*) and between a solid surface and the fluid itself (*skin friction*). These effects are formalized through a **macroscopic balance**, leading to the *Equation of mechanical energy*

$$l = -\int_1^2 \frac{dp}{\rho_f} - g\Delta z - \frac{\Delta \langle u \rangle^2}{2} - \frac{\Delta p}{\rho_f} \tag{3.7}$$

In this way, one can deduce the specific mechanical power $l$ (related to fluid density $\rho_f$, pressure $p$, gravity acceleration $g$, height $z$, and bulk velocity $\langle u \rangle$) to provide by means of a mechanical pump, depending on the friction amounting to the pressure drop or specific *load loss* $\Delta p / \rho_f$: a measure of the inherent irreversibility.

In some flow cases, the velocity profiles and other related results can be computed based on Newton's Law of viscosity and using a *balance of forces*. The pressure drop and the effect of gravity can be at stake in the flow assessment.

### 3.2.1  Internal Flow

Let us start with the common situation of **fluid flows within a limiting surface** or conduit, as in simple fluid displacement and in unit operations: the flow can be moved by a fan in the fluid is a gas, or a pump if it is a liquid.

### 3.2.1.1 Flow Through a Tube

In this case, one can deduce on $\Delta p$ by writing a balance of forces on a fluid cylinder element, having a radius $r$ and a length $dx$, subject to **uniform linear** (steady) **motion** along the coordinate $x$, in a round tube of radius $R$ (Fig. 3.5, top). We will demonstrate that, in said assumptions, the velocity profile depends solely on $r$; the elemental pressure drop $dp$ will be determined by the (tangential) wall shear stress (resistance to motion) that occurs to the element's surfaces.

Let us consider the elemental fluid $CV$ of Fig. 3.5, top, and let us apply the balance of forces[11] as shown in Fig. 3.5, bottom:

$$p\pi r^2 - (p + dp)\pi r^2 - \tau_{rx} 2\pi r dx = 0 \tag{3.8}$$

Simplifying, one can see that the pressure gradient along the tube is given by such tangential shear stress:

$$-\frac{dp}{dx} = \frac{2}{r}\tau_{rx} \tag{3.9}$$

For this geometry, with the Newton's Law of viscosity

$$\tau_{yx} = -\mu\frac{\partial u}{\partial y} \tag{3.5}$$

one derives

$$\frac{dp}{dx} = \frac{2\mu}{r}\frac{du}{dr} \tag{3.10}$$

Here, the two terms on the left- and right-hand sides depend separately on $x$ and $r$. The only way to satisfy this equality is that the two terms must be equal to a same constant:

**Fig. 3.5** Tube flow. Top: geometry nomenclature; bottom: location of vectors (represented by white solid arrows) for the balance of forces

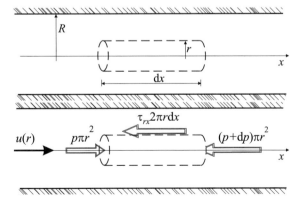

---

[11]This balance is valid regardless of the inherent fluid regime.

$$\frac{dp}{dx} = \frac{2\mu}{r}\frac{du}{dr} = \text{const} \tag{3.11}$$

Carrying the integration of the left-hand side term, along the tube's length $L$:

$$p(L) - p(0) = -\Delta p = L \times \text{cost} \Rightarrow \frac{dp}{dx} = -\frac{\Delta p}{L} \tag{3.12}$$

We see that the **pressure drop is constant**: pressure $p$ varies linearly along the tube. Now, carrying the integration of the right-hand side term in Eq. (3.11), between 0 and $R$, we get a *parabolic velocity profile*:

$$u(r) = \frac{R^2}{4\mu}\left(-\frac{dp}{dx}\right)\left[1 - \frac{r^2}{R^2}\right] \tag{3.13}$$

In the half plane $r^+$, $u$ decreases with $r$, canceling out for $r = R$.[12] The situation depicted by Eq. (3.13) is called a *fully developed flow*. This situation can be assumed when the location along $x$ is far enough from tube inlet or fluid dynamic disturbance occurrences.[13] We already reported on this laminar velocity profile, in Fig. 3.1. The maximum velocity is reached on the tube axis ($r = 0$):

$$u_{max} = \frac{R^2}{4\mu}\left(-\frac{dp}{dx}\right) \tag{3.14}$$

from which we learn that, besides the pressure drop, the motion uniformity implies that the **maximum velocity is constant**, as well. We can write Eq. (3.14) as

$$-\frac{dp}{dx} = \frac{4\mu u_{max}}{R^2}$$

therefore carrying the integration one obtains:

$$-(p_2 - p_1) = -\Delta p = \left(\frac{4\mu u_{max}}{R^2}\right)(x_2 - x_1) = \frac{4\mu L u_{max}}{R^2} \tag{3.15}$$

with $L$ is the tube length subject to the pressure drop.

In many cases, the flow rate is at stake. The definition for the *nominal* **mass flow rate**, for a flow passing through the section $\Omega$, is

$$\dot{m} \equiv \rho_f \langle u \rangle \Omega \tag{3.16}$$

---

[12]This is the *adherence* or "*no-slip*" BC: the fluid must adhere to the solid surface, meaning that its velocity relative to it is zero. In other words, the force of attraction between the fluid particles and solid particles (adhesive forces) is greater than that between the fluid particles (cohesive forces). However, applications exist when a "*slip*" BC must be assumed instead: in this case, the fluid relative velocity is nonzero.

[13]In tube flows, is conventionally assumed that this distance is some 20 internal diameters.

using the bulk space-**averaged velocity** defined in this case as

$$\langle u \rangle \equiv \frac{1}{\Omega} \int_\Omega u \, d\Omega = \frac{1}{\pi R^2} \int_0^R u(r) 2\pi r \, dr \qquad (3.17)$$

We can drop in for $u(r)$ what is gathered from Eq. (3.13):

$$\langle u \rangle = -\frac{1}{\pi R^2} \left( \frac{R^2}{4\mu} \frac{dp}{dx} \pi r^2 \Big|_0^{R^2} - \frac{R^2}{4\mu} \frac{dp}{dx} \frac{\pi r^4}{2R^2} \Big|_0^{R^2} \right) = -\frac{dp}{dx} \left( \frac{R^2}{4\mu} - \frac{R^2}{8\mu} \right)$$

Developing the algebra:

$$\langle u \rangle = \frac{R^2}{8\mu} \left( -\frac{dp}{dx} \right) = \frac{1}{2} u_{max} \qquad (3.18)$$

and recalling the result of Eq. (3.14), we see that the mass flow rate can be conveniently calculated using the value of maximum velocity determined at tube centerline.[14] Furthermore, Eq. (3.13) can also be rewritten in terms of $\langle u \rangle$:

$$u(r) = 2\langle u \rangle \left[ 1 - \frac{r^2}{R^2} \right] \qquad (3.19)$$

From the knowledge of the mass flow rate $\dot{m}$, Eq. (3.15) can be recast as[15]

$$-\Delta p = \frac{8\mu L}{\pi \rho_f R^4} \dot{m} \qquad (3.20)$$

We see that the pressure drop increases along the tube ($p_2 < p_1$), and when using a smaller tube with the same mass flow rate (which means length $L$ and maximum velocity $u_{max}$), one has a greater pressure drop.

Now, we are also interested in plotting the velocity and the tangential shear stress along the tube section. Equation (3.13) shows that the velocity distribution is parabolic with $r$. Equation (3.15) yields $u_{max}$, at tube centerline, once the pressure drop is known:

---

[14]The local flow velocity can be measured (along with its temporal variations) at a given point, purposely located in the flow stream, by means of an instrument called a velocity *anemometer*, based on some kind of signal transduction. Hot-wire, laser-Doppler (from Austrian physicist C. DOPPLER in mid-nineteenth century), ultrasonic and acoustic resonance anemometers are common. Pressure anemometers can also be used depending on the application: for example, by using a *pitot tube* (invented by the French engineer H. PITOT in the early eighteenth century). Some device of this sort is relatively insensitive to rapid velocity fluctuations, so to perform velocity-averaging over several seconds.

[15]Also called the *Hagen–Poiseuille equation of laminar flow in a tube*, as proposed independently by German civil engineer G.H.L. HAGEN and French physiologist J.L.M. POISEUILLE in the early nineteenth century. This equation does not hold at tube's entrance.

**Fig. 3.6** Tube flow. Left:
parabolic velocity
distribution $u(r)$; right:
linear momentum flux
distribution $\tau_{rx}(r)$

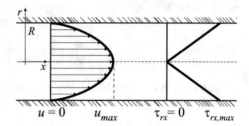

$$u_{max} = \frac{-\Delta p R^2}{4\mu L} \tag{3.21}$$

Similarly, the distribution of the momentum flux, that is, the tangential shear stress $\tau_{rx}$, can be obtained by manipulation of Eq. (3.9): integrating by variable separation, a linear function with $r$ is obtained:

$$- \Delta p = \frac{2\tau_{rx} L}{r} \tag{3.22}$$

with the maximum value falling on the tube walls:

$$\tau_w \equiv \tau_{rx,max} = \left(\frac{-\Delta p}{2L}\right) R \tag{3.23}$$

or, using Eq. (3.21)

$$\tau_w = 4\mu \frac{\langle u \rangle}{R} \tag{3.24}$$

These distribution are presented in Fig. 3.6.

### 3.2.1.2   Flow Along Parallel Plates

In an analogous way, the flow limited by two plates spaced by a distance $L$ undergoes to a fully developed flow far enough from inlet. With $x$ the streamwise coordinate, one derives in the same way than with Eq. (3.11) that

$$\frac{dp}{dx} = \frac{\mu}{y} \frac{du}{dy} = \text{const} \tag{3.25}$$

Carrying the integration of the right-hand side term between 0 and $L/2$, we get a parabolic velocity profile:

$$\boxed{u(y) = \frac{3}{2} \langle u \rangle \left[ 1 - \left(\frac{y}{L/2}\right)^2 \right]} \tag{3.26}$$

with

$$\langle u \rangle = \frac{L^2}{12\mu}\left(-\frac{\mathrm{d}p}{\mathrm{d}x}\right) \tag{3.27}$$

and

$$u_{\max} = 1.5\langle u \rangle \tag{3.28}$$

### 3.2.1.3 Comparison Between Flow Regimes in Tubes

It is useful to note the similarities and differences in the velocity distribution and load loss relationships for fully developed laminar or turbulent flows. In Sect. 3.1.1 (p. 67), it was seen that laminar flow is an orderly fluid aggregation, and as such, it can be described completely by an analytical method, i.e., by integrating the differential equation that governs the phenomenon, as performed in the present section. For turbulent flow, on the other hand, the velocity fluctuates with time chaotically at each point in the $CV$, so that the analytical method seen so far cannot help, except for the macroscopic balance Eq. (3.7) (that can be applied just as well in turbulent regimes, for it refers to time-averaged values of velocity). These aspects will be examined in more detail later in Sect. 3.4 (p. 108).

So in order to compare the various relationships, we must distinguish between (bulk) space- $\langle u \rangle$ and time-averaging $\overline{u}$. We recall that, for a laminar flow (Re < 2100), it is:

$$u(r) = \left[1 - \left(\frac{r}{R}\right)^2\right]u_{\max} \tag{3.29}$$

$$\langle u \rangle = \frac{1}{2}u_{\max} \tag{3.30}$$

$$-\Delta p = \frac{8}{\pi}\frac{\mu L}{\rho_{\mathrm{f}}R^4}\dot{m}, \quad \text{linear dependence on } \dot{m} \tag{3.31}$$

For turbulent flow, when $10^4 < \mathrm{Re} < 10^5$, the following empirical relationships are valid instead [3]:

$$\overline{u(r)} \approx \left(1 - \frac{r}{R}\right)^{1/7}\overline{u_{\max}} \tag{3.32}$$

$$\langle \overline{u} \rangle \approx \frac{4}{5}\overline{u_{\max}} \tag{3.33}$$

$$-\Delta p \approx 0.0198\left(\frac{2}{\pi}\right)^{7/4}\frac{\mu^{1/4}L}{\rho_{\mathrm{f}}R^{19/4}}\dot{m}^{7/4}, \quad \text{nonlinear dependence on } \dot{m} \tag{3.34}$$

We already reported on this turbulent velocity profile of Eq. (3.32), flatter at the centerline and much steeper at the wall, in Fig. 3.1.

### 3.2.1.4  Friction Factors and Pressure Drop

Fully developed flow in tubes and ducts indeed reports that the longitudinal velocity $u$ and the wall shear stress $\tau_w$ are not function of the streamwise position. It is customary to express the wall shear stress in dimensionless form, by using the purposely *Fanning friction factor* $f$[16]

$$\boxed{f \equiv \frac{\tau_w}{\frac{1}{2}\rho_f \langle u \rangle^2}} \tag{3.35}$$

For *fully developed flow in tubes*, combining this definition with Eq. (3.24) and Reynolds number definition Eq. (3.2)

$$\mathrm{Re}_D \equiv \frac{\langle u \rangle D}{\nu} \tag{3.2 revisited}$$

we have

$$f = \frac{16}{\mathrm{Re}_D} \tag{3.36}$$

Now, sections other than round are found in the applications, such as the previous parallel plates duct and rectangular channels. To generalize the notations, the *hydraulic diameter* $D_H$ of an arbitrary tube section $\Omega$ is employed based on the internal *wetted perimeter* $P$, as

$$D_H \equiv \frac{4\Omega}{P} \tag{3.37}$$

With Eq. (3.36), a variety of $f\,\mathrm{Re}_{D_H}$ product values exist, for cross-sectional shapes other that round, as grouped in Table 3.3.

It is worth noting that the Fanning friction factor is associated with the *Darcy friction factor* $f_D$[17] which is employed in friction calculations for general configurations, even in the presence of turbulence and surface roughness. The value of $f_D$ can be obtained, when the Reynolds number of Eq. (3.2) and the relative roughness

---

[16] After the American engineer J.T. FANNING at the end of nineteenth century.

[17] As proposed by French engineer H.P.G. DARCY in mid-nineteenth century. It is $f_D \equiv 4f$. The Darcy friction factor $f_D$ is related to the specific dissipation term $\Delta p/\rho_f$ in the following version of the equation of mechanical energy for open systems:

$$l = -\int_1^2 \frac{dp}{\rho_f} - g\Delta z - \frac{\Delta \langle u \rangle^2}{2} - \frac{\Delta p}{\rho_f} \tag{3.7}$$

through the phenomenological *Equation of Darcy–Weisbach* (with J.L. WEISBACH a German mathematician and engineer operating in the nineteenth century) relating the head loss due to friction $\Delta p$ along a given length $L$ of tube of diameter $D$ to the bulk flow velocity $\langle u \rangle$ for an incompressible fluid:

$$\frac{\Delta p}{L} = f_D \frac{\rho_f}{D} \frac{\langle u \rangle^2}{2}.$$

**Table 3.3** Hydraulic diameter $D_H$ and $f\mathrm{Re}_{D_H}$ products of Eq. (3.36) for some cross-sectional shape [4]

| Cross-sectional shape | $D_H$ | $f\mathrm{Re}_{D_H}$ |
|---|---|---|
| Round | $D$ | 16 |
| Equilateral triangle, side $L$ | $0.66L$ | 13.3 |
| Square, side $L$ | $0.25L$ | 14.2 |
| Rectangular, long side $4L$, short side $L$ | $0.4L$ | 18.3 |
| Wide parallel plates | $\rightarrow 0.5$ | 24 |

$\varepsilon/D$ are known, on the ordinate of the *Moody's chart*,[18] as illustrated in Fig. 3.7. Recalling from Thermodynamics, this chart can be generated in the turbulent regime by using $f_D = f_D(\mathrm{Re}_D, \varepsilon/D)$ formulations, such as Chen's Equation.[19] Indicative values of the equivalent roughness $\varepsilon$ for new tubes, for a variety of materials, are listed in Table 3.4.

## 3.2.2 External Flow

External flows occur in a great deal of situations as well: a flow over a model in a wind tunnel, for example, is moved by the tunnel fan. Alternatively, the surface may move through a steady fluid, as in an airplane flight. Another case is the **liquid film in free fall along an inclined plate**. This situation is found in flows along wet walls, evaporation and absorption of gas fractions, coatings in a variety of biotech processes. The starting point is again the *balance of forces* on a fluid elements flowing **in laminar regime and in uniform** (steady) **motion** along the $z$ coordinate, along the plate for the $CV$ depicted in Fig. 3.8.

It is first assumed that the film thickness $\delta$ (along $x$) is small with respect the film width (along $y$) and length $L$ (along $z$) of plate. Consequently, due to the system extension along $y$, the velocity component $v$ can be neglected. Furthermore, due to the adopted regime, the velocity component $u$ can also be neglected, and if no flow perturbations (such as detachment from plate) are allowed, for limited mass flow rates the viscous forces will prevail to the film acceleration, so that $w$ becomes soon constant with $z$. Therefore, $w$ will solely depend on film depth (along $x$), and all

---

[18] As elaborated by the American engineer and professor L.F. MOODY in mid-twentieth century.

[19] As proposed by Chen, N.H.: An Explicit Equation for Friction Factor in Pipe. Industrial & Engineering Chemistry Fundamental (1979) 18 3:

$$\frac{1}{\sqrt{f_D}} = 2.0 \log \left\{ \frac{1}{3.7065} \frac{\varepsilon}{D} - \frac{5.0452}{\mathrm{Re}_D} \log \left[ \frac{1}{2.8257} \left( \frac{\varepsilon}{D} \right)^{1.1098} + \frac{5.8506}{\mathrm{Re}_D^{0.8981}} \right] \right\}$$

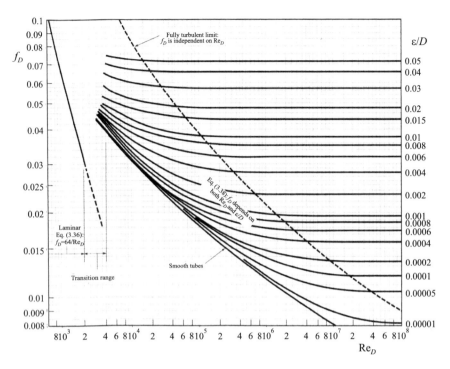

**Fig. 3.7**  Moody's chart for *Darcy friction factor* $f_D = f_D(\mathrm{Re}_D, \varepsilon/D)$

**Table 3.4**  Equivalent
roughness $\varepsilon$ for new tubes or
channels in various materials

| Material | $\varepsilon$ (mm) |
|---|---|
| Concrete | 0.3–3.0 |
| Drawn tubing | 0.0015 |
| Glass | 0.0 (smooth) |
| Iron, cast | 0.26 |
| Iron, galvanized | 0.15 |
| Iron, wrought | 0.045 |
| Plastic | 0.0 (smooth) |
| Steel, commercial | 0.045 |
| Steel, riveted | 0.9–9.0 |
| Wood stave | 0.18-0.9 |

**Fig. 3.8** Geometry
nomenclature for the case of
a liquid flow that rises in a
tank, forming a free surface,
then flowing as a film over
and down along the inclined
plate

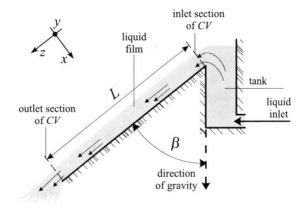

velocity components are qualified:

$$u = 0, \quad v = 0, \quad w = w(x) \tag{3.38}$$

It can be noted that nothing is changing along $y$ and $z$; therefore, even pressure $p$ is solely function on $x$: the elemental pressure drop $dp$ will be given by the tangential stress (resistance to motion) applied to the surfaces of normal $x$. Lastly, viscosity $\mu$ and density $\rho_f$ are taken as constant.

**Fig. 3.9** Film flowing along
an inclined plate:
nomenclature of fluid
element

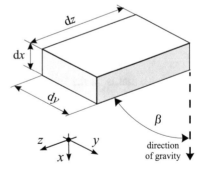

**Fig. 3.10**  Film flowing along an inclined plate: vectors placement for the balance of forces

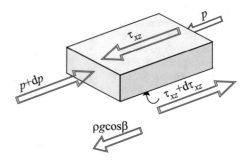

The fluid element at stake is reported in Fig. 3.9, having $dx$, $dy$, and $dz$ dimensions. Let us apply the balance of forces per unit surface, over the fluid element, as well as the one per unit volume, as in Fig. 3.10:

$$pdxdy - (p + dp)dxdy + \tau_{xz}dydz - (\tau_{xz} + d\tau_{xz})dydz + \rho_f gdxdydz \cos \beta = 0$$
$$(3.39)$$

Simplifying:

$$dpdxdy - d\tau_{xz}dydz + \rho_f gdxdydz \cos \beta = 0 \qquad (3.40)$$

Due to *film thinness*, it is practically $dp = 0$ for which

$$\frac{d\tau_{xz}}{dx} = \rho_f g \cos \beta \qquad (3.41)$$

Separating variables as usual and integrating a first time:

$$\tau_{xz} = \rho_f g \cos \beta x + a \qquad (3.42)$$

For $x = 0$, that is, at the free surface, it is $\tau_{xz} = a = 0$. Therefore:

$$\tau_{xz} = \rho_f g \cos \beta x \qquad (3.43)$$

We can use Newton's Law of viscosity

$$\tau_{yx} = -\mu \frac{\partial u}{\partial y} \qquad (3.5)$$

for the interpretation of $\tau_{xz}$, so that

$$\tau_{xz} \equiv -\mu dw/dx \qquad (3.44)$$

then yielding

$$dw = -\frac{1}{\mu}\tau_{xz}dx \qquad (3.45)$$

Now, if we drop in Eq. (3.43), we get

$$w(x) = -\frac{\rho_f g \cos \beta}{2\mu} x^2 + b \tag{3.46}$$

which is the general integral for $w(x)$. Let us proceed by part integration: a first boundary condition is obtained by exploiting the *no-slip* BC (see Note 12), prescribing that close to the plate (for $x = \delta$) one has $w = 0$. Therefore,

$$b = \frac{\rho_f g \cos \beta \delta^2}{2\mu} \tag{3.47}$$

can be substituted back in the general integral Eq. (3.46), yielding, after some algebra manipulation, a *parabolic velocity profile*:

$$w(x) = \frac{\rho_f g \delta^2 \cos \beta}{2\mu} \left[ 1 - \left( \frac{x}{\delta} \right)^2 \right] \tag{3.48}$$

The maximum velocity is reached at film free surface (for $x = 0$) so that

$$w_{max} = \frac{\rho_f g \delta^2 \cos \beta}{2\mu} \tag{3.49}$$

Again, it is useful to compute the average velocity $\langle w \rangle$ leading to the nominal mass flow rate $\dot{m}$. Equation (3.17) is still valid as applied to the $z$-direction:

$$\langle w \rangle \equiv \frac{1}{\delta \times 1} \int_0^\delta w(x) \mathrm{d}x \tag{3.50}$$

Thickness $\delta$ is constant and can be introduced in the integral expression, in order to get $\mathrm{d}(x/\delta)$ as the integration argument. Then, we get:

$$\langle w \rangle = \int_0^\delta w(x) \mathrm{d}\left( \frac{x}{\delta} \right) = \frac{\rho_f g \delta^2 \cos \beta}{2\mu} \int_0^\delta \left[ 1 - \left( \frac{x}{\delta} \right)^2 \right] \mathrm{d}\left( \frac{x}{\delta} \right)$$

Integrating:

$$\langle w \rangle = \frac{\rho_f g \delta^2 \cos \beta}{2\mu} \left( 1 - \frac{1}{3} \right) = \frac{\rho_f g \delta^2 \cos \beta}{3\mu} = \frac{2}{3} w_{max} \tag{3.51}$$

In Fig. 3.11, we see the progress of velocity distribution $w$, which is parabolic with $x$, with the maximum value $w_{max}$ at the free surface dictated by Eq. (3.49), as well as the progress of the moment flux distribution $\tau_{xz}$, which is linear with $x$.

**Fig. 3.11** Film flowing along an inclined plate. Top right: velocity distribution; bottom left: momentum flux distribution

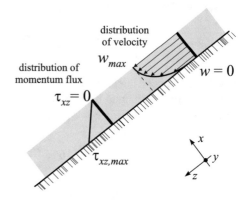

### 3.2.3   Summary of the Solution of 1-D Problems

So, in case of laminar flow having only one nonzero velocity component, the employed sequence is as follows:

1. the nonzero velocity component is identified, along with the coordinate on which it depends;
2. the differential balance of forces (or momentum flux) is written on an infinitesimal $CV$;
3. the differential equation is integrated, with its BCs, yielding the momentum distribution;
4. Newton's Law of viscosity is exploited to get the velocity distribution;
5. this distribution is used to calculate other quantities of interest such as the maximum and bulk average velocities, and the force exerted on the solid surfaces.

A list of the most common boundary conditions follows:

- *at the impermeable solid–fluid interface*, the velocity components (tangential and normal to the interface) are equal to those of the solid surface: *No-slip condition*
- *at the fluid–fluid interface*, i.e., for a plane at constant $x$, the tangential components $v$ and $w$ are continuous through the plane, as well as all stresses: *Slip condition*
- *at the permeable* (porous) *solid–fluid interface*, a velocity component, the normal one to the interface, is continuous through the interface: *Selected slip condition*
- *at the liquid–gas interface*, as for a plane at constant $x$, the stresses $\tau_{xy}$ and $\tau_{xz}$ are zero (for small variations of the relative velocity, which is likely as the viscosity of gases is much lesser than those of liquids)

In the development of these elementary viscous flow cases, no mass transfer has been considered at the interface, such as adsorption, absorption, dissolution, evaporation/condensation, melting, nor chemical reactions of sorts.

## 3.3 Fundamental Equations of Fluid Mechanics

In order to obtain the velocity and pressure distributions in arbitrary geometries, it is necessary to resort to a *generalized formulation* to determine the distribution of velocity vector **w**, through its components $u$, $v$, and $w$. This formulation can be applied, through the mass and momentum balances, and solved by using numerical methods, as outlined later in Sect. 3.6 (p. 120).

Let us consider first a laminar flow field, represented by its streamlines[20] through an infinitesimal control volume $CV$ fixed in a Cartesian 2-D space and confined by its control surface $CS$, as represented in Fig. 3.12.

### *3.3.1 Equation of Continuity*

Then, let us apply the *mass flow rate*

$$\langle u \rangle \equiv \frac{1}{\pi R^2} \int_0^R u(r) 2\pi r \, dr \tag{3.17}$$

in a mass balance on the $CV$, in terms of components of through the $CS$:

$$\boxed{\dot{m} \text{ entering } CS} =$$

$$\boxed{\dot{m} \text{ exiting } CS} + \boxed{\text{variation of } m \text{ in } CV, \text{ in the unit } \theta} \tag{3.52}$$

**Fig. 3.12** The infinitesimal control volume $CV$ and its confining control surface $CS$, crossed by a fluid flow (streamlines) in its reference frame, with indication of the vector composition of **w**. Note 18 of Chap. 2 applies

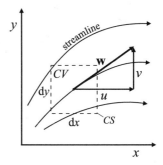

---

[20]Streamlines are a family of curves that are instantaneously tangent to the velocity vector describing the flow.

**Fig. 3.13** Mass balance
components through the $CS$

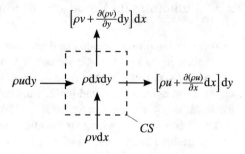

Similarly to what showed with

$$\dot{Q}_x \, (x + \mathrm{d}x) \equiv \dot{Q}_x \, (x) + \frac{\partial \dot{Q}_x \, (x)}{\partial x} \mathrm{d}x \qquad (2.20)$$

this balance can be carried out by means of the Taylor's series expansion, getting the components depicted in Fig. 3.13. Summing them up, we get:

$$\rho_{\mathrm{f}} u \mathrm{d}y + \rho_{\mathrm{f}} v \mathrm{d}x = \left[ \rho_{\mathrm{f}} u + \frac{\partial \, (\rho_{\mathrm{f}} u)}{\partial x} \mathrm{d}x \right] \mathrm{d}y + \left[ \rho_{\mathrm{f}} v + \frac{\partial \, (\rho_{\mathrm{f}} v)}{\partial y} \mathrm{d}y \right] \mathrm{d}x + \frac{\partial}{\partial \theta} \, (\rho_{\mathrm{f}} \mathrm{d}x \mathrm{d}y)$$
$$(3.53)$$

Simplifying, we get:

$$\boxed{\frac{\partial \rho_{\mathrm{f}}}{\partial \theta} + \frac{\partial \, (\rho_{\mathrm{f}} u)}{\partial x} + \frac{\partial \, (\rho_{\mathrm{f}} v)}{\partial y} = 0} \qquad (3.54)$$

which is the **Equation of continuity**[21] *in dimensional Cartesian form*. In a 3-D space, an appropriate third term (the $z$ derivative of $\rho_{\mathrm{f}} w$) will show up.

Other forms of Eq. (3.54) may be encountered, by using the *vector-tensor nota-tion*[22] or the *vector notation*. In this last case, if we unfold the derivatives in Eq. (3.54), we have:

$$\frac{\partial \rho_{\mathrm{f}}}{\partial \theta} + u \frac{\partial \rho_{\mathrm{f}}}{\partial x} + v \frac{\partial \rho_{\mathrm{f}}}{\partial y} + \rho_{\mathrm{f}} \left( \frac{\partial u}{\partial x} + \frac{\partial v}{\partial y} \right) = 0 \qquad (3.55)$$

---

[21] Also called the *Equation of conservation of mass*.

[22] The Equation of continuity reads

$$\frac{\partial \rho_{\mathrm{f}}}{\partial \theta} + \frac{\partial \left( \rho_{\mathrm{f}} w_j \right)}{\partial x_j} = 0$$

Subscript $j$ takes the values rotating from 1 to 3 (corresponding to Cartesian directions), with the implicit meaning that the terms with repeated subscript must be summed up together.

or, using the divergence operator[23]

$$\boxed{\frac{D\rho_f}{D\theta} + \rho_f (\nabla \cdot \mathbf{w}) = 0}$$
(3.56)

Here, the concept of **material derivative** has been employed, to take into full account the variation of scalars or vectors in a flow field. [24]

By looking at Eq. (3.55), for those flows in which the *temporal and spatial varia-tions in density are negligible*, with respect to local variations in velocity,[25] then the Equation of continuity reads

$$\boxed{\frac{\partial u}{\partial x} + \frac{\partial v}{\partial y} = 0}$$
(3.57)

### 3.3.2  Equation of Momentum Conservation

Next, let us stick with the balance of momentum on the $CV$, provided that all *pertinent forces*, along with their sign, are accounted for. The projection of the force definition

$$\sum \mathbf{F} = \frac{d(m\mathbf{w})}{d\theta}$$
(3.1)

---

[23] When $\nabla$ is applied to a vector by a scalar product (with reference to Note 13 of Chap. 2), gives its *divergence* (i.e., a scalar). When using the Cartesian 3-D coordinates:

$$\nabla \cdot \mathbf{w} \equiv \frac{\partial u}{\partial x} + \frac{\partial v}{\partial y} + \frac{\partial w}{\partial z}$$

It is worth to recall the *divergence or Gauss's theorem* (after German mathematician J.C.F. GAUSS, in the early nineteenth century) that states that the volume integral of the divergence of a vector field $\mathbf{F}$ inside a $CV$ is equal to the net flux of $\mathbf{F}$ through the $CS$:

$$\int_{CV} (\nabla \cdot \mathbf{F}) dV = \int_{CS} (\mathbf{F} \cdot \mathbf{n}) dS.$$

[24] The material derivative $D/D\theta$ is the sum of the *local* (time-dependent) and the *convective* terms. When applied to scalars, such as $\rho_f$, it yields:

$$\frac{D\rho_f}{D\theta} = \text{first 3 terms of Eq. (3.55)}$$

By the way, when applied to vectors, such as $\mathbf{w}$ itself, it yields:

$$\frac{D\mathbf{w}}{D\theta} = \frac{\partial \mathbf{w}}{\partial \theta} + \mathbf{w} \cdot \nabla \mathbf{w} = \frac{\partial \mathbf{w}}{\partial \theta} + u \frac{\partial u}{\partial x} + v \frac{\partial v}{\partial y}.$$

[25] In these conditions, the flow is said to be *incompressible*.

when performed along $x$, gives:[26]

$$\boxed{\sum F_x \text{ applied on } CV} = \boxed{\text{variation of } mu \text{ in } CV, \text{ in the unit } \theta} \qquad (3.58)$$

On the left-hand side, the relevant forces are subdivided into:

- *contact* or *surface forces*, due to the action of the surrounding fluid or solid surface on the $CS$ encasing the $CV$ (giving rise to fluid or skin friction). We can generalize the definition of stress

$$\frac{F}{A} = \tau_{yx} \propto \frac{W}{L} \qquad (3.4)$$

  as these forces can be subdivided in *normal stresses* $\sigma$ and *tangential stresses* $\tau$. When projected on coordinate axes, Note 7 applies for their nomenclature;
- *mass forces*, due to contingent force fields applied on the $CV$. Following the development of the energy conservation of Sect. 2.3.1 (p. 26), we duly note that the mass force is an *extrinsic volumetric source/sink term for the momentum flux*, accounting for every generative or dissipative effect, such as gravity or magnetohydrodynamics. An important case is the **buoyancy force (variation of fluid density) due to fluid temperature gradient**, as will explained later in Chap. 4. Without digging in its consistence for the time being, and being a vector, this term will be denoted by **S** and, or sake of simplicity, it will be taken as *uniform across the CV*; note that **it can be positive or negative, depending on the case at stake**.

This balance can again be carried out by means of the Taylor's series expansion. So with reference to Fig. 3.14 and the *term on the left-hand side* of Eq. (3.58), let us sum up these terms with their appropriate sign, as applied to $CS$ and $CV$, respectively:

$$- \sigma_{xx} dy + \left( \sigma_{xx} + \frac{\partial \sigma_{xx}}{\partial x} dx \right) dy - \tau_{yx} dx + \left( \tau_{yx} + \frac{\partial \tau_{yx}}{\partial y} dy \right) dx \pm S_x dx dy$$

that is

$$\left( \frac{\partial \sigma_{xx}}{\partial x} \right) dx dy + \left( \frac{\partial \tau_{yx}}{\partial y} \right) dx dy \pm S_x dx dy \qquad (3.59)$$

From now on, a positive sign of source term **S** components represents a positive contribution of momentum to the $CV$.

---

[26]Equation (3.58) is a particular form of the general *Cauchy's momentum Equation* $\frac{1}{\rho_f} \nabla \cdot \boldsymbol{\sigma} + \mathbf{S} =$ D**w**/D$\theta$, as proposed by the earlier mentioned A.- L. CAUCHY in the early nineteenth century, the Equation describes the momentum transport in continua.

**Fig. 3.14** Force balance
components at the $CS$ and
$CV$, along $x$

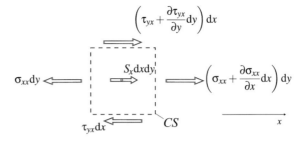

As for the *term on the right-hand side* of Eq. (3.58), the $CV$ is commonly fixed
and the **fluid is incompressible**[27] such that:

$$\frac{D(mu)}{D\theta} = m\frac{D(u)}{D\theta} = m\left(\frac{\partial u}{\partial \theta} + \mathbf{w}\cdot\nabla u\right) \qquad (3.60)$$

In the *steady state*, the only term left is the convective one:

$$m\frac{D(u)}{D\theta} = m\left(u\frac{\partial u}{\partial x} + v\frac{\partial u}{\partial y}\right) \qquad (3.61)$$

The two sides of Eq. (3.58) can be put together:

$$\left(\frac{\partial \sigma_{xx}}{\partial x}\right)dxdy + \left(\frac{\partial \tau_{yx}}{\partial y}\right)dxdy \pm S_x dxdy = m\left(u\frac{\partial u}{\partial x} + v\frac{\partial u}{\partial y}\right)$$

Simplifying and carrying on the balance of forces also along $y$ in the same fashion
(see Fig. 3.15), we derive:

$$\rho_f\left(u\frac{\partial u}{\partial x} + v\frac{\partial u}{\partial y}\right) = \frac{\partial \sigma_{xx}}{\partial x} + \frac{\partial \tau_{yx}}{\partial y} \pm S_x \qquad (3.62a)$$

$$\rho_f\left(u\frac{\partial v}{\partial x} + v\frac{\partial v}{\partial y}\right) = \frac{\partial \sigma_{yy}}{\partial y} + \frac{\partial \tau_{xy}}{\partial x} \pm S_y \qquad (3.62b)$$

---

[27]That is, density $\rho_f$ is constant. This assumption is very frequently verified in the applications,
specially when multiphysics is involved. Adding to what specified with Note 25, fluid incom-
pressibility does not imply that the flow itself is incompressible. Even compressible fluids can—to
good approximation—realize an incompressible flow. Incompressible flow implies that the density
remains constant within a parcel of fluid that moves with the flow velocity.

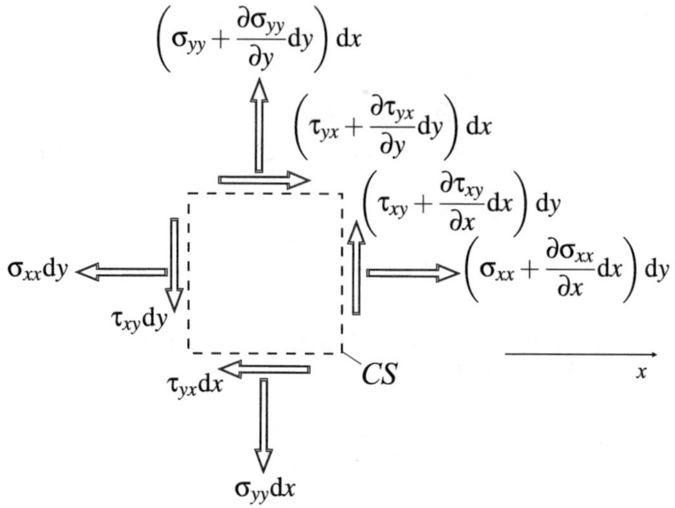

$$\left(\sigma_{yy} + \frac{\partial \sigma_{yy}}{\partial y}dy\right)dx$$

$$\left(\tau_{yx} + \frac{\partial \tau_{yx}}{\partial y}dy\right)dx$$

$$\left(\tau_{xy} + \frac{\partial \tau_{xy}}{\partial x}dx\right)dy$$

$$\left(\sigma_{xx} + \frac{\partial \sigma_{xx}}{\partial x}dx\right)dy$$

$\sigma_{xx}dy$

$\tau_{xy}dy$

$\tau_{yx}dx$   *CS*

$x$

$\tau_{yx}dx$

$\sigma_{yy}dx$

**Fig. 3.15**  Contact force balance at the $CS$ and $CV$, along $x$ and $y$. Let us note the $\tau$ symmetry, to ensure the $CV$ equilibrium, in the sense of the continuum mechanics

In the *unsteady state*, we simply have:

$$\rho_f\left(\frac{\partial u}{\partial \theta} + u\frac{\partial u}{\partial x} + v\frac{\partial u}{\partial y}\right) = \frac{\partial \sigma_{xx}}{\partial x} + \frac{\partial \tau_{yx}}{\partial y} \pm S_x \tag{3.63a}$$

$$\rho_f\left(\frac{\partial v}{\partial \theta} + u\frac{\partial v}{\partial x} + v\frac{\partial v}{\partial y}\right) = \frac{\partial \sigma_{yy}}{\partial y} + \frac{\partial \tau_{xy}}{\partial x} \pm S_y \tag{3.63b}$$

### 3.3.3  Generalized Constitutive Relationships

In order to resolve the stresses included in Eqs. (3.62) or (3.63) and come up with the pressure and the velocity components, it is necessary now to resort to the stress–velocity gradient constitutive relationships or **Stokes' Viscosity Law**,[28] which adequately generalizes Newton's Law

$$\tau_{yx} = -\mu\frac{\partial u}{\partial y} \tag{3.5}$$

The subject is treated by the vast field of fluid dynamics; therefore, we will include here a simplified scrutiny, only. It is:

---

[28] As proposed by Irish mathematician and physicist G.G. STOKES, in mid-nineteenth century.

$$\sigma_{xx} = -p + 2\mu\frac{\partial u}{\partial x} \qquad (3.64a)$$

$$\sigma_{yy} = -p + 2\mu\frac{\partial v}{\partial y} \qquad (3.64b)$$

$$\tau_{xy} = \tau_{yx} = \mu\left(\frac{\partial u}{\partial y} + \frac{\partial v}{\partial x}\right) \qquad (3.64c)$$

Equations (3.64) were derived in the same assumption of incompressible fluid which lead to Eq. (3.60).[29] A generalized form of Eqs. (3.64) in tensorial notation for *compressible fluids* may also be encountered, involving the concept of a further material property: the *expansion viscosity* $\mu'$. The 9 stress components can be written as follows:

$$\tau_{ij}^{v} = \mu\left(\frac{\partial w_i}{\partial x_j} + \frac{\partial w_j}{\partial x_i}\right) - \left[p + \left(\frac{2}{3}\mu - \mu'\right)\frac{\partial w_k}{\partial x_k}\right]\delta_{ij} \qquad (3.65)$$

with $\delta_{ij}$ the Kronecker delta,[30] which is 1 if $i = j$ and 0 otherwise. The superscript v is to remind that we are dealing with a laminar flow. With Note 22 in mind, the subscript rotation is now extended to all three subscripts. Also, it is still $\tau_{ij} = \tau_{ji}$ for $i \neq j$. A generalized form of Eqs. (3.64) in vector-tensor notation is also common.[31]

Let us take the two stress terms in Eq. (3.62a); we can take Eqs. (3.64a, 3.64c) so that we derive for a *constant dynamic viscosity fluid*:

---

[29] In this case (when $\nabla \cdot \mathbf{w} = 0$, i.e., a liquid), the pressure appears as the negative average of the normal stresses:

$$p = -\frac{\sigma_{xx} + \sigma_{yy}}{2}$$

while for a fluid at rest (as in hydrostatics) we have simply $p = -\sigma_{xx} = -\sigma_{yy}$.

[30] Named after German mathematician L. KRONECKER, mid-eighteenth century.

[31] Even more concisely:

$$\boldsymbol{\tau}^{v} = \mu\left[\nabla\mathbf{w} + (\nabla\mathbf{w})^{\mathrm{T}}\right] - \left[p + \left(\frac{2}{3}\mu - \mu'\right)(\nabla \cdot \mathbf{w})\right]\boldsymbol{\delta}$$

in which $\nabla\mathbf{w}$ is the velocity gradient tensor (with components $\partial w_j/\partial x_i$), $(\nabla\mathbf{w})^{\mathrm{T}}$ is its transpose (with components $\partial w_i/\partial x_j$), and $\boldsymbol{\delta}$ is the unit tensor (with components $\delta_{ij}$). Sometimes, in lieu of $\boldsymbol{\delta}$, the identity matrix $\mathbf{I}$ is employed.

Note that, when performing multiplications in vector-tensor notation, different parentheses indicate the type of result produced: ( ), [ ], or { } yield a scalar (zeroth-order tensor), a vector (first-order tensor), or a tensor proper (second-order tensor), respectively.

Finally, it has to be considered that, regardless of the value of $\mu'$ (e.g., 0 for monoatomic gases, and 3 and 100 times larger than $\mu$ for water and benzene, respectively), the effect of expansion or compression on the flow (when the value of $(\nabla \cdot \mathbf{w})$ is positive or negative) is usually very small, except in special cases: when density changes are induced over very short scales, such as molecular distances (e.g., in the interior of shock waves) or infinitesimal durations (e.g., in high-intensity ultrasound).

$$\frac{\partial \sigma_{xx}}{\partial x} + \frac{\partial \tau_{yx}}{\partial y} = \left( -\frac{\partial p}{\partial x} + 2\mu \frac{\partial^2 u}{\partial x^2} \right) + \mu \left( \frac{\partial^2 u}{\partial y^2} + \frac{\partial^2 v}{\partial y \partial x} \right) \qquad (3.66)$$

Cleaning this up, the *right-hand side* can be written as:

$$= -\frac{\partial p}{\partial x} + \mu \left( \frac{\partial^2 u}{\partial x^2} + \frac{\partial^2 u}{\partial y^2} \right) + \mu \left( \frac{\partial^2 u}{\partial x^2} + \frac{\partial^2 v}{\partial y \partial x} \right), \text{ or}$$

$$= -\frac{\partial p}{\partial x} + \mu \left( \frac{\partial^2 u}{\partial x^2} + \frac{\partial^2 u}{\partial y^2} \right) + \mu \frac{\partial}{\partial x} \left( \frac{\partial u}{\partial x} + \frac{\partial v}{\partial y} \right) \qquad (3.67)$$

Based on the continuity Eq. (3.57), last term of Eq. (3.67) is dropped and Eq. (3.66) is simplified as follows:

$$\frac{\partial \sigma_{xx}}{\partial x} + \frac{\partial \tau_{yx}}{\partial y} = -\frac{\partial p}{\partial x} + \mu \left( \frac{\partial^2 u}{\partial x^2} + \frac{\partial^2 u}{\partial y^2} \right) \qquad (3.68)$$

### 3.3.4 Navier–Stokes Equations in Rectangular Coordinates

With no further complications, a similar development can be carried over along $y$. At this point, using these results, Eqs. (3.62) can be substituted by the following equation of momentum conservation:

$$\rho_f \left( u \frac{\partial u}{\partial x} + v \frac{\partial u}{\partial y} \right) = \mu \left( \frac{\partial^2 u}{\partial x^2} + \frac{\partial^2 u}{\partial y^2} \right) - \frac{\partial p}{\partial x} \pm S_x \qquad (3.69a)$$

$$\rho_f \left( u \frac{\partial v}{\partial x} + v \frac{\partial v}{\partial y} \right) = \mu \left( \frac{\partial^2 v}{\partial x^2} + \frac{\partial^2 v}{\partial y^2} \right) - \frac{\partial p}{\partial y} \pm S_y \qquad (3.69b)$$

Together with the independent variables $u$ and $v$, in coherence with the findings of the balance of forces in Sect. 3.2 (p. 74), we acknowledge that a second independent variable $p$, **the pressure of the fluid, must now be taken into account**.

As we join Eqs. (3.69) with the continuity Eq. (3.57), we come up with the **Navier–Stokes Equations**,[32] or the **fundamental Equations of fluid mechanics**, in the assumption of *fluid with constant rheological properties*.[33] In this form, the Navier–Stokes Equations Eqs. (3.57, 3.69) are valid for laminar flows, whereas in case of turbulent flows some modifications are due, as will be carried out later.

In the *unsteady state* [3], we have in vector notation:

---

[32] As proposed by French engineer and physicist C.L.M.H. NAVIER in the early nineteenth century.
[33] It is $d\rho_f / d\theta = 0$.

$$\frac{\partial \rho_f}{\partial \theta} + (\nabla \cdot (\rho_f \mathbf{w})) = 0 \tag{3.70a}$$

$$\underbrace{\rho_f \left[ \frac{\partial \mathbf{w}}{\partial \theta} + (\mathbf{w} \cdot \nabla) \mathbf{w} \right]}_{\textcircled{1}} = \underbrace{\mu \nabla^2 \mathbf{w}}_{\textcircled{2}} \underbrace{- \nabla p}_{\textcircled{3}} \underbrace{\pm \mathbf{S}}_{\textcircled{4}} \tag{3.70b}$$

Other forms of Eq. (3.70b) may be encountered.[34]

Let us keep our focus on Eq. (3.70b). The various terms appearing here all hold the meaning of forces, *in competition* in every flow point: term ① is the *inertia force* (subdivided in *unsteady acceleration* and *convective acceleration*), term ② is the *viscous force*, term ③ is the *pressure force*, and ④ is the *source/sink* term such as the *mass force*.

Recalling the mechanism of heat transfer by conduction, we find some **mathematical and physical analogies**, and some **differences** as well: *the viscous term is analogous to the diffusion term*, and the source/sink terms are present as well (although with different physical meaning. But **two new concepts arise now: the convective acceleration in ① and the gradient of pressure ③**, which are truly the *distinctive terms* for this transport mechanism.

*The system of Eqs.* (3.70), together with the associate initial conditions (stating the velocity and pressure distribution in the $CV$ at the beginning of the process), and with the appropriate number of BCs onto the $CS$, *allows one to simultaneously*[35] *determine the functions* $\mathbf{w}(x, y, \theta)$ and $p(x, y, \theta)$, in the specified assumption of constant dynamic viscosity. The possible BCs can subdivided in *velocity type* or *pressure type*, whereas mixed types should not be exercised onto the same $CS$ to avoid improperly posed problems. Usually, 4 kinds of BC can be employed:

1. the most common BC that can occur resembles the first-kind BC for temperature seen earlier at the end of Sect. 2.3.1 (p. 26): the fluid enters the $CV$ with a specified velocity value or distribution and/or adheres to the solid surface so it attains its velocity. For a stationary surface, we get the *no-slip* BC; when the surface is moving, we get a *slip* BC.

---

[34] As an example, with Note 27 in mind, for variable properties, steady state flows it is:

$$(\mathbf{w} \cdot \nabla)\mathbf{w} = \nabla \cdot \nu \left[ \nabla \mathbf{w} + (\nabla \mathbf{w})^{\mathrm{T}} \right]$$

while recalling Eq. (3.65) and using the vector-tensor notation:

$$\rho_f \frac{D\mathbf{w}}{D\theta} = (\nabla \cdot \tau^{\mathrm{v}}) \pm \mathbf{S}.$$

[35] The solutions for velocity and pressure clearly depend one another.

2. far from the perturbing surface, the velocity approaches some prescribed value: the *undisturbed velocity* BC. Very often, a second-kind BC will be use, by declaring the first derivative of some component of velocity zero at that boundary. Backflow condition can also be dealt with, by requesting such derivative to be deduced from the interior values of velocity;
3. in case of porous surface, suction, injection, or discharge can be prescribed so that generally a *normal velocity component* BC is applied;
4. a pressure type BC is encountered whenever an interface between two fluids occurs, such as the *free surface* BC. Across the interface, the pressure in each fluid is the same, except when the *surface tension* is important: in this case, the pressure difference is related to the surface tension. In addition to the pressure condition, the component of velocity normal to the interface is the same in both fluids.

It is evident that the system of Eq. (3.70) cannot be integrated with analytical techniques, except in few simplistic cases: that is why we turn then to computational methods, later in Sect. 3.6 (p. 120).

### 3.3.5  Particular Cases

- In some situation of unbounded flows or when the effect of viscous force or shear ② can be deemed negligible (such as the airflow around and far from airplane wings and the ocean currents), with no source/sink term Eq. (3.70b) simplifies as

$$\rho_f \left[ \frac{\partial \mathbf{w}}{\partial \theta} + (\mathbf{w} \cdot \nabla) \, \mathbf{w} \right] = -\nabla p \qquad (3.71)$$

which is known as the *Euler equation* for *inviscid flows*.[36]
- In contrast, when the viscous shear effects is so large as to balance the driving effect of the pressure, that is, if the Reynolds number defined by Eq. (3.2) is *very small*, the effect of inertia ① is negligible and Eq. (3.70b) simplifies as

$$\nabla p = \mu \nabla^2 \mathbf{w} \qquad (3.72)$$

which is valid for "creeping" or *Stokes' flows*,[37] such as glaciers and cold molasses.
- Generally, there are flows which are driven by pressure or gravitational forces, in which the bulk flow is *irrotational*, that is, a flow in which each fluid element may accelerate or deform but not rotate. One reason for rotation of flow is the no-slip

---

[36] As proposed by Swiss mathematician and physicist L. EULER in mid-eighteenth century.
[37] Named after the aforementioned scientist G.G. STOKES.

condition due to flow by a solid surface (effect of viscous force); therefore, these flows are implicitly inviscid. An irrotational flow is identified by[38]

$$\nabla \times \mathbf{w} = 0 \tag{3.73}$$

This means that the velocity field $\mathbf{w}$ is a conservative vector field given by the gradient of a *scalar potential function* $\varphi$, that is,

$$\mathbf{w} = \nabla \varphi \tag{3.74}$$

In terms of $\varphi$, recalling Note 13 of Chap. 2, the Equation of continuity

$$\frac{D\rho_f}{D\theta} + \rho_f \nabla \cdot \mathbf{w} = 0 \tag{3.56}$$

becomes in steady state

$$\frac{\partial^2 \varphi}{\partial x^2} + \frac{\partial^2 \varphi}{\partial y^2} + \frac{\partial^2 \varphi}{\partial z^2} = 0 \tag{3.75}$$

which is the laplacian of $\varphi$ (see Note 24 of Chap. 2). This case is called a *potential flow*. A possible use of $\varphi$ is the solution of a flow field, where a potential flow solution (outside the boundary layer) is coupled to a solution of the boundary layer itself. Another use of $\varphi$ is its exploitation in 2-D flow solution. Here, a *stream function* $\psi(x, y)$ can be defined that is constant along a streamline (Fig. 3.12). Therefore, $\psi$ would be constant along a solid surface limiting the flow. Now, let us use the Equation of continuity

$$\frac{\partial u}{\partial x} + \frac{\partial v}{\partial y} = 0 \tag{3.57}$$

to define

$$u \equiv \frac{\partial \psi}{\partial y}, \quad v \equiv -\frac{\partial \psi}{\partial x} \tag{3.76}$$

In case of irrotational flow, while the $x$- and $y$- components ($\xi$ and $\eta$) are identically zero, it is for the $z$-component $\zeta \equiv \frac{\partial v}{\partial x} - \frac{\partial u}{\partial y} = 0$ (see Note 38), or

$$\frac{\partial^2 \psi}{\partial x^2} + \frac{\partial^2 \psi}{\partial y^2} = 0 \tag{3.77}$$

---

[38]When $\nabla$ is applied to a vector by a vector product (with reference to Note 13 of Chap. 2), it gives its *curl* (i.e., a vector). When using the Cartesian 3-D coordinates:

$$\nabla \times \mathbf{w} \equiv \left( \frac{\partial w}{\partial y} - \frac{\partial v}{\partial z} \right) \mathbf{i} + \left( \frac{\partial u}{\partial z} - \frac{\partial w}{\partial x} \right) \mathbf{j} + \left( \frac{\partial v}{\partial x} - \frac{\partial u}{\partial y} \right) \mathbf{k}.$$

**Table 3.5** Comparison between viscous flow in a tube and 1-D cylindrical conduction

| Occurrence | Viscous tube flow | Cylindrical conduction |
|---|---|---|
| First integration gives the distribution of | $\tau_{rz}$ | $\dot{q}_r$ |
| Second integration gives the distribution of | $v_z$ | $T - T_0$ |
| Boundary condition for $r = 0$ is | $\tau_{rz} = 0$ | $\dot{q}_r = 0$ |
| Boundary condition for $r = r_0$ is | $v_z = 0$ | $T - T_0 = 0$ |
| Transport property | $\mu$ | $\lambda$ |
| Source term | $(p - p_0)/L$ | $\dot{e}'''$ |
| Assumptions on medium | $\mu$ = const. | $\lambda, \lambda_e$ = const. |

Hence, the stream function also satisfies Laplace's equation (see Note 32 of Chap. 2) for an irrotational flow.

- For fluid flowing *through a porous medium*, the seepage/filtration/superficial *velocity is related to the applied pressure difference*. This means that the governing momentum Equation (3.70b) can be replaced, for an isotropic medium, by the *Darcy's Equation*[39]

$$\mathbf{w} = -\frac{\kappa}{\mu} \nabla p \qquad (3.78)$$

with $\kappa$ the *permeability* of the medium, depending only on its geometry; $[\kappa]$=m$^2$, while a frequent derived unit is the darcy D (with 1 D=1 $\mu$m$^2$). For an incompressible fluid, the continuity Equation is again

$$(\nabla \cdot \mathbf{w}) = 0 \qquad (3.57 \text{ revisited})$$

so when we combine Eqs. (3.57, 3.78), we find that *the flow in porous media is a potential flow* governed by a relationship like Eq. (3.75).

### 3.3.6 Other Coordinate Systems

A useful analogy between viscous flow and conduction heat transfer in cylindrical geometries can be readily set up in Table 3.5, with respect to variables during integration and material properties. In the same assumptions as Eqs. (3.70), the Navier–Stokes Equations for the cylindrical geometry *in scalar notation* (that is, for the geometry illustrated in Fig. 2.12) are reported here, for the sake of completeness.

---

[39] As proposed by the aforementioned H.P.G. DARCY in its simplest form. More complex alternatives exist, such as the Forchheimer and Brinkman variations.

$$\text{continuity}: \quad \frac{\partial \rho_f}{\partial \theta} + \frac{1}{r}\frac{\partial \rho_f r v_r}{\partial r} + \frac{1}{r}\frac{\partial \rho_f v_\phi}{\partial \phi} + \frac{\partial \rho_f v_z}{\partial z} = 0 \tag{3.79a}$$

$$\text{along } r: \quad \rho_f \left( \frac{\partial v_r}{\partial \theta} + v_r \frac{\partial v_r}{\partial r} + \frac{v_\phi}{r}\frac{\partial v_r}{\partial \phi} + v_z \frac{\partial v_r}{\partial z} - \frac{v_\phi^2}{r} \right)$$
$$= \mu \left\{ \frac{\partial}{\partial r}\left[ \frac{1}{r}\frac{\partial}{\partial r}(r v_r) \right] + \frac{1}{r^2}\frac{\partial^2 v_r}{\partial \phi^2} + \frac{\partial^2 v_r}{\partial z^2} - \frac{2}{r^2}\frac{\partial v_\phi}{\partial \phi} \right\} \tag{3.79b}$$
$$- \frac{\partial p}{\partial r} + S_r$$

$$\text{along } \phi: \quad \rho_f \left( \frac{\partial v_\phi}{\partial \theta} + v_r \frac{\partial v_\phi}{\partial r} + \frac{v_\phi}{r}\frac{\partial v_\phi}{\partial \phi} + v_z \frac{\partial v_\phi}{\partial z} - \frac{v_r v_\phi}{r} \right)$$
$$= \mu \left\{ \frac{\partial}{\partial r}\left[ \frac{1}{r}\frac{\partial}{\partial r}(r v_\phi) \right] + \frac{1}{r^2}\frac{\partial^2 v_\phi}{\partial \phi^2} + \frac{\partial^2 v_\phi}{\partial z^2} - \frac{2}{r^2}\frac{\partial v_r}{\partial \phi} \right\} \tag{3.79c}$$
$$- \frac{1}{r}\frac{\partial p}{\partial \phi} + S_\phi$$

$$\text{along } z: \quad \rho_f \left( \frac{\partial v_z}{\partial \theta} + v_r \frac{\partial v_z}{\partial r} + \frac{v_\phi}{r}\frac{\partial v_z}{\partial \phi} + v_z \frac{\partial v_z}{\partial z} \right)$$
$$= \mu \left[ \frac{1}{r}\frac{\partial}{\partial r}\left( r \frac{\partial v_z}{\partial r} \right) + \frac{1}{r^2}\frac{\partial^2 v_z}{\partial \phi^2} + \frac{\partial^2 v_z}{\partial z^2} \right] - \frac{\partial p}{\partial z} + S_z \tag{3.79d}$$

The Navier–Stokes Equations in a spherical coordinate system (that is, for the geometry in figure at right in Note 12) are much less frequently found [3].

## 3.3.7 Turbulent Flow

Turbulent flows are encountered in many practical applications. We seized the opportunity already, in Sect. 3.1.1 (p. 67), to illustrate briefly the flow regimes and the onset of turbulence. We also compared, for a 1-D tube flow, the results relative to laminar or bulk turbulent regime at the end of Sect. 3.2.1 (p. 74). Turbulence is also very important as it often underlies heat and mass transfer in flow systems. Qualitatively, the exchange mechanism in turbulent flow can be pictured as a magnification of the molecular exchange in laminar flow. No simplified treatment as the present one would even superficially scratch the vast field of study, so the following material serves as an introduction to popular descriptive and solution methods.

### 3.3.7.1 Turbulence Features

In general, all of the following apply to turbulent flows:

- turbulent flows are *fluctuating*, even in the steady state;

- turbulent flows are *inherently 3-D* (or 4-D, if we consider their transient nature). The mean values of velocity may depend on two coordinates only, but the instantaneous field appears essentially random; moreover, the *velocity components are generally correlated* themselves;
- turbulent flows contain a *great deal of vorticity* or rotation. Stretching of vortices is one of the principal mechanisms by which the "intensity" of turbulence is increased. Turbulent flows may also contain *coherent structures*, i.e., deterministic, repeatable patterns that are responsible for a large part of the mixing. Among these, *vortices and eddies* (swirling and reversing flow) are common;
- fluctuations increase the rate at which transported quantities (momentum, energy, species) are subject to *mixing* and cause this quantities to fluctuate as well, i.e., when parcels of fluid having differing values of at least one of these quantities are brought into contact;
- due to this mixing, the reduction of the velocity gradients due to the action of viscosity yields a decrease of the kinetic energy of the flow: *mixing is a dissipative process*, meaning that some of the flow energy is irreversibly converted into internal energy of the fluid;
- turbulent flows may fluctuate over a broad range of length and timescales, making their virtualization a difficult task indeed.

The effect produced by turbulence may be desirable or not, depending on the application. Intense mixing is indeed useful when heat and/or mass transfer is needed in processing; both of these may be augmented by some OoM. On the other hand, increased mixing of momentum leads to greater frictional forces, so that more flow energy is demanded by the external mechanical source/sink, and stronger drag results to a bounding surface or an immersed body.

Comprehension and prediction of turbulence are essential to achieve good design, system verification, and outcomes control. The study of turbulence has been primarily an experimental art, but with the increase of computation potential and in the presence of almost impossible measurements for some variables in certain configurations, modeling and simulation play now an important role.


### 3.3.7.2   High-End Turbulence Modeling Methods

The most accurate approach to evaluate turbulence flows is to solve all of the motion details by the Navier–Stokes Equations. In this way, the computed flow field is equivalent to a realization of an experimental laboratory flow run. Called the *Direct Numerical Simulation* (DNS), this approach requires special and costly treatments spacewise and timewise, as all of the spatial and temporal gradients must be solved. Firstly, it is necessary that all of the significant structures of the turbulence and all of the kinetic energy dissipation have been captured. This twofold requirement on the *numerical space grid* dictates the size on the $CV$ over which the computation must be performed (the largest turbulent eddy) and at the same time the size of the grid to employ that must be the smallest scale at which the viscosity is active,[40] especially near the solid confining surfaces where solid/fluid interaction and hence momentum transfer occur. Secondly, the computational accuracy requires the adoption of a small *numerical time grid*, as well as the storage of flow history at each flow point, to ensure physically sound simulations. Then, it is clear that the application of DNS requires huge computational resources so that it is viable for limited Reynolds number flows, only.

A compromise between DNS and simpler methods, the Large Eddy Simulation (LES), is based on the assumption that the relevant scales in turbulent flows can be subdivided into large scale and small scale. The former is generally more energetic and is the main contributor to the momentum transport. LES therefore attempts to solve rigorously these large-scale motions, while the small-scale motions are believed to be more universal in character and are modeled empirically. Although LES is defined on a 4-D basis, as well, it is much less costly and more flexible than DNS, so it is a preferred method for flows in which the Reynolds number is too high or the geometry is too complex to be feasibly handled by DNS.

Usually, the engineering analysis requests the macroscopically observed *mean* values rather than every microscopic detail. To this end, the solution of the averaged Navier–Stokes Equations for turbulent flow would suffice, so that a viable starting point in this elementary presentation is the averaging applied to the Navier–Stokes Equations for unsteady state laminar flow. Then, we need to turn to a suitable transport model of turbulence that will be deployed with similar forms than the governing equations developed so far.

### 3.3.7.3  Reynolds-Averaged Navier–Stokes Equations

If we measure the flow velocity in one point in a given $CV$, we may observe some irregular, rapid fluctuation about a mean value (Fig. 3.16). Let us consider the governing Equation

$$\rho_f \left[ \frac{\partial \mathbf{w}}{\partial \theta} + (\mathbf{w} \cdot \nabla)\mathbf{w} \right] = \mu \nabla^2 \mathbf{w} - \nabla p \pm \mathbf{S} \qquad (3.70b)$$

The actual value for a component of velocity $\mathbf{w}$ may be regarded as the sum of the mean (time-averaged) value and its fluctuation:[41]

$$u(\theta) \equiv \overline{u} + u' \qquad (3.80)$$

The mean value can be obtained, over a $\Delta\theta$ interval large enough from the initial time $\theta_0$, by letting

$$\overline{u}(\theta) \equiv \frac{1}{\Delta\theta} \int_{\theta_0}^{\theta_0 + \Delta\theta} u(\theta)\mathrm{d}\theta \qquad (3.81)$$

---

[40]Also called the Kolmogorov scale, after Russian mathematician A. KOLMOGOROV, in mid-twentieth century.

[41]Also called *the Reynolds decomposition*.

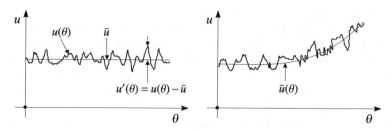

**Fig. 3.16** Velocity measurement in one flow point, with indication of fluctuations around an average value. At left: a steadily driven turbulent flow (with constant $\bar{u}$); at right, a situation in which $\bar{u}$ depends on time

The same applies for pressure $p$ and the source/sink term **S**. Also, alternate averaging (or filtering) could be performed, in the statistical or spatial sense.

According to the nature of fluctuations, some *properties on averaging* apply: double-averaging is equivalent to single-averaging, and the average of derivation is equivalent to the derivative of the average. Therefore, the following relations are true:

$$\overline{u'} = 0, \quad \overline{\bar{u}} = \bar{u}, \quad \overline{\bar{u}u'} = 0, \quad \overline{\frac{\partial u}{\partial \theta}} = \frac{\partial \bar{u}}{\partial \theta}, \quad \overline{\frac{\partial u}{\partial x}} = \frac{\partial \bar{u}}{\partial x} \qquad (3.82)$$

The quantity $\overline{u'^2} = \overline{u'u'}$ will not, however, be zero, and in fact, the ratio $\sqrt{\overline{u'^2}}/\langle \bar{u} \rangle$ can be taken as a turbulent reference, or *turbulence intensity* along $x$.

Quantities such as $\overline{u'v'}$ are also nonzero. The reason for this is that the local motions in the $x$- and $y$-directions are *correlated*: the fluctuation in one direction is not independent of the fluctuation in the other direction. These averaged products of fluctuating properties have an important role in turbulent transfer, for any of the mechanism (momentum, heat and mass) considered.

At this point, let us take the Navier–Stokes Equations, by replacing **w** by its equivalent $\overline{\mathbf{w}} + \mathbf{w}'$, $p$ by its equivalent $\overline{p} + p'$, and **S** by its equivalent $\overline{\mathbf{S}} + \mathbf{S}'$ [5]. For a 3-D space obtained from the extension along $z$ of the one depicted in Fig. 3.12, the incompressible version of the Equation of continuity

$$(\nabla \cdot (\rho_f \mathbf{w})) = 0 \qquad (3.70)$$

becomes:

$$\frac{\partial}{\partial x} \left( \bar{u} + u' \right) + \frac{\partial}{\partial y} \left( \bar{v} + v' \right) + \frac{\partial}{\partial z} \left( \bar{w} + w' \right) = 0 \qquad (3.83)$$

while the $x$-component of the Equation of momentum conservation Eq. (3.70b) reads as follows (without the source/sink term, for the sake of simplicity):

$$\rho_f \frac{\partial}{\partial \theta} \left( \bar{u} + u' \right) +$$

$$\rho_f \left[ \frac{\partial}{\partial x} \left( \bar{u} + u' \right) \left( \bar{u} + u' \right) + \frac{\partial}{\partial y} \left( \bar{v} + v' \right) \left( \bar{u} + u' \right) + \frac{\partial}{\partial z} \left( \bar{w} + w' \right) \left( \bar{u} + u' \right) \right] =$$

$$\mu \nabla^2 \left( \bar{u} + u' \right) - \frac{\partial}{\partial x} \left( \bar{p} + p' \right) \pm \left( \bar{S}_x + S'_x \right) \tag{3.84}$$

Now, *if we take the average of Eqs. (3.83, 3.84), we obtain* the **Reynolds-averaged** (time-smoothed) *Equations of continuity* and of $x$-momentum conservation, part of the **Reynolds-averaged Navier–Stokes Equations** (RANS) (which will include the momentum conservation along the other directions), in the assumption of *fluid with constant rheological properties*:

$$\frac{\partial \bar{u}}{\partial x} + \frac{\partial \bar{v}}{\partial y} + \frac{\partial \bar{w}}{\partial z} = 0 \tag{3.85}$$

$$\rho_f \left( \frac{\partial \bar{u}}{\partial \theta} + \frac{\partial \bar{u}\,\bar{u}}{\partial x} + \frac{\partial \bar{u}\,\bar{v}}{\partial y} + \frac{\partial \bar{u}\,\bar{w}}{\partial z} \right) + \rho_f \underbrace{\left( \frac{\partial \overline{u'u'}}{\partial x} + \frac{\partial \overline{u'v'}}{\partial y} + \frac{\partial \overline{u'w'}}{\partial z} \right)}_{\text{turb. mom. transport due to fluctuations}} = \tag{3.86}$$

$$\mu \nabla^2 \bar{u} - \frac{\partial \bar{p}}{\partial x} \pm \bar{S}_x$$

Relationships similar to Eq. (3.86) occur along $y$ and $z$. If we compare Eqs. (3.85, 3.86) with Eqs. (3.57, 3.69a), we see that

- the Equation of continuity is the same as we had previously, except that **w** components are now replaced by $\bar{\mathbf{w}}$ components;
- the Equation of momentum conservation now has $\bar{\mathbf{w}}$ components and $\bar{p}$ where we previously had **w** components and $p$. In addition to the material derivative of the $\bar{\mathbf{w}}$ components, the under-braced terms (and the similar ones along $y$ and $z$) appear, which represent **the momentum transport associated with the turbulent fluctuations, which adds to the molecular transport**.

Other notations of RANS may be encountered, to emphasize on the difference between the *turbulent moment flux* tensor $\boldsymbol{\tau}^t$ (whose components are usually referred to as the *Reynolds stresses*) and the *time-averaged viscous moment flux* tensor $\overline{\boldsymbol{\tau}^v}$,[42] in vector-tensor notations and no source/sink term:

---

[42] $\boldsymbol{\tau}^t$ has the following components:

$$\tau^t_{xx} \equiv \rho_f \overline{u'u'}, \quad \tau^t_{xy} \equiv \rho_f \overline{u'v'}, \quad \text{and so on, or} \quad \tau^t_{ij} \equiv \rho_f \overline{w'_i w'_j}$$

In analogy with Eqs. (3.64, 3.65), $\overline{\boldsymbol{\tau}^v}$ has the following components:

$$\bar{\tau}_{xx} \equiv 2\mu \frac{\partial \bar{u}}{\partial x}, \quad \bar{\tau}_{xy} \equiv \mu \left( \frac{\partial \bar{u}}{\partial y} + \frac{\partial \bar{v}}{\partial x} \right), \quad \text{and so on, or} \quad \bar{\tau}_{ij} \equiv \mu \left( \frac{\partial \bar{w}_i}{\partial x_j} + \frac{\partial \bar{w}_j}{\partial x_i} \right).$$

$$(\nabla \cdot \overline{\mathbf{w}}) = 0 \tag{3.87a}$$

$$\rho_f \frac{D\overline{\mathbf{w}}}{D\theta} = \left(\nabla \cdot \left[\overline{\boldsymbol{\tau}^{v}} + \boldsymbol{\tau}^{t}\right]\right) \tag{3.87b}$$

or, for *variable rheological properties*

$$\frac{\partial (\rho_f \overline{w}_i)}{\partial x_i} = 0 \tag{3.88a}$$

$$\frac{\partial (\rho_f \overline{w}_i)}{\partial \theta} + \frac{\partial}{\partial x_j} \left(\rho_f \overline{w}_i \overline{w}_j\right) = \frac{\partial}{\partial x_j} \left(\overline{\tau_{ij}^{v}} + \tau_{ij}^{t}\right) \tag{3.88b}$$

### 3.3.7.4   The Momentum Eddy Diffusivity

Another common notation exploits the concept of *momentum eddy diffusivity* $\varepsilon_v$ (with $[\varepsilon_v]$=m²/s), so to evidence the added contribution of turbulence to molecular viscosity. Defining the *apparent turbulent shear stress*

$$\frac{1}{\rho_f} \tau_{ij}^{t} \equiv -\overline{w_i' w_j'} \equiv \varepsilon_v \frac{\partial \overline{w}_i}{\partial x_j} \tag{3.89}$$

RANS Eq. (3.88) becomes, in the assumption of an *incompressible fluid*:

$$\frac{\partial \overline{w}_i}{\partial \theta} + \frac{\partial}{\partial x_j} \left(\overline{w}_i \overline{w}_j\right) = \frac{\partial}{\partial x_j} \left(v_t \frac{\partial \overline{w}_i}{\partial x_j}\right) - \frac{1}{\rho_f} \frac{\partial \overline{p}}{\partial x_i} \pm \overline{S}_i \tag{3.90}$$

through an incremented *turbulent kinematic viscosity* $v_t$:

$$v_t = v + \varepsilon_v \tag{3.91}$$

To make a reasonable estimate of $\varepsilon_v$, we can use a simple *OoM analysis*. Imagine a parcel of fluid situated at a distance $y$ from the wall in a given turbulent flow: its average flowwise velocity being $\overline{u}(x, y, z)$. If this packet moves toward the wall based on an eddy having diameter $l$, until a new position $y - l$, its average velocity will become $\overline{u}(x, y - l, z)$. The distance $l$ is called the *mixing length* over which the packet maintains its original nature:[43] the velocity fluctuation will be of the following OoM[44]

---

[43] As proposed by German physicist L. PRANDTL in the early nineteenth century. See the reference for Turbulence recalled later in Further Reading.

[44] Sign $\sim$ means "is of the same OoM as."

$$|u'| \sim \overline{u}(x, y, z) - \overline{u}(x, y - l, z) \sim l \frac{\partial \overline{u}}{\partial y}$$

Due to the 3-D nature of the pushing eddy, it can be argued that the other velocity components are of the same OoM as $u'$. Therefore, substituting any of these values into the definition of $\varepsilon_v$ Eq. (3.89), we obtain

$$\varepsilon_v \sim l^2 \left| \frac{\partial \overline{u}}{\partial y} \right| \tag{3.92}$$

Measurements of $\overline{u}$ along $y$ suggest that the mixing length $l$ is proportional to the distance $y$ itself:

$$l = \kappa y \tag{3.93}$$

where $\kappa \cong 0.4$ is called the von Kármán's constant,[45] and

$$\varepsilon_v = \kappa^2 y^2 \left| \frac{\partial \overline{u}}{\partial y} \right| \tag{3.94}$$

Unlike molecular viscosity, which is a parameter of the fluid, the momentum eddy diffusivity $\varepsilon_v$ depends strongly on the position within the flow, while going to zero at the flow-limiting wall. Therefore, in practical flow solution, more viable approaches are preferred.

Going back to the Reynolds stresses, $\boldsymbol{\tau}^t$ or $\tau_{ij}^t$, they are not related in a simple way, as are their viscous counterparts, $\overline{\boldsymbol{\tau}^v}$ or $\overline{\tau_{ij}^v}$, to the gradients of the average velocity. Instead, they depend on the position in the flow and the turbulence intensity. To express these fluxes is the task of a turbulent model, such as the simplest one based on empirical relationships, that is revisited next.

### 3.3.7.5 A Base-Line Turbulent Model

With laminar flows, energy dissipation and transport of mass and momentum normal to flow streamlines are mediated by viscosity. With turbulent flows, we have seen in Eq. (3.86) that the effect of turbulence to momentum transfer can be represented by an increased viscosity. This notion has been enforced by the *Boussinesq eddy viscosity hypothesis*[46] that has a limited complexity in implementation and may provide reasonable results for many flows. Recalling the definition of shear stress for comparison,

$$\tau_{ij}^v = \mu \left( \frac{\partial w_i}{\partial x_j} + \frac{\partial w_j}{\partial x_i} \right) - \left[ p + \left( \frac{2}{3} \mu - \mu' \right) \frac{\partial w_k}{\partial x_k} \right] \delta_{ij} \tag{3.65}$$

---

[45] After Hungarian-American mathematician and engineer T. VON KÁRMÁN, in mid-twentieth century.

[46] After French physicist and mathematician V.J. BOUSSINESQ, at beginning twentieth century.

this conjecture leads to the *eddy viscosity model* that will translate in describing the increased momentum transport associated with the turbulent fluctuations of Eqs. (3.86) with a turbulent moment flux tensor:

$$\rho_f \overline{w_i' w_j'} \equiv \mu_t \left( \frac{\partial \overline{w}_i}{\partial x_j} + \frac{\partial \overline{w}_j}{\partial x_i} \right) - \frac{2}{3} \rho_f k \delta_{ij} \tag{3.95}$$

whereas $\mu_t$ is the *turbulence eddy viscosity*, and $k$ is the *turbulent kinetic energy* $\frac{1}{2} \overline{w_i' w_i'}$.

In its simplest description, such as the case at stake here, turbulence is characterized by a *velocity scale k* and a *length scale L*. The difficulty in prescribing the turbulence quantities suggests that one might use PDEs to compute them. So, a model which derives the needed quantities from two such equations is a logical choice.

1. Generally, with a PDE describing the balance over the turbulent kinetic energy $k$ such as the following (for a *fluid with variable rheological properties*), the velocity scale can be determined:

$$\boxed{\frac{\partial (\rho_f k)}{\partial \theta} + \frac{\partial \left( \rho_f \overline{w}_j k \right)}{\partial x_j} = \frac{\partial}{\partial x_j} \left[ \left( \mu + \frac{\mu_t}{\sigma_k} \right) \frac{\partial k}{\partial x_j} \right] - \rho_f \varepsilon + S_k} \tag{3.96}$$

where $\varepsilon$ is called the *turbulent dissipation*, and the source/sink term $S_k$, the rate of production of turbulent energy by the mean flow, is computed from

$$S_k \equiv -\rho_f \overline{w_i' w_j'} \frac{\partial \overline{w}_i}{\partial x_j} \tag{3.97}$$

2. So, an additional variable $\varepsilon$ does show up in Eq. (3.96). Thus, in lieu of solving for a length scale, another equation for $\varepsilon$ is required to achieve model closure:

$$\boxed{\frac{\partial (\rho_f \varepsilon)}{\partial \theta} + \frac{\partial \left( \rho_f \overline{w}_j \varepsilon \right)}{\partial x_j} = \frac{\partial}{\partial x_j} \left[ \left( \mu + \frac{\mu_t}{\sigma_\varepsilon} \right) \frac{\partial \varepsilon}{\partial x_j} \right] + S_\varepsilon} \tag{3.98}$$

where the source/sink term $S_\varepsilon$, the rate of production of dissipation, is empirically computed from given constants:

$$S_\varepsilon \equiv C_{1\varepsilon} S_k \frac{\varepsilon}{k} - C_{2\varepsilon} \rho_f \frac{\varepsilon^2}{k} \tag{3.99}$$

In this model, the turbulence eddy viscosity is expressed as

$$\mu_t \equiv \rho_f C_\mu \frac{k^2}{\varepsilon} \tag{3.100}$$

while the following model constant can be successfully adopted in most cases:

$$C_{1\varepsilon} \equiv 1.44, \quad C_{2\varepsilon} \equiv 1.92, \quad C_{\mu} \equiv 0.09, \quad \sigma_k \equiv 1.0, \quad \sigma_\varepsilon \equiv 1.3 \quad (3.101)$$

Clearly, Eqs. (3.96, 3.98) are perfectly suited to be solved in the transport mechanism framework specified so far, following the general transport PDE for variable $\phi$ introduced with Eq. (1.1). With the notations deployed so far, RANS Eq. (3.90) can be rewritten as:

$$\rho_f \frac{\partial \overline{w}_i}{\partial \theta} + \rho_f \frac{\partial}{\partial x_j} \left( \overline{w}_i \overline{w}_j \right) = \frac{\partial}{\partial x_j} \left[ (\mu + \mu_t) \frac{\partial \overline{w}_i}{\partial x_j} \right] - \frac{\partial \overline{p}}{\partial x_i} \pm \overline{S}_i \qquad (3.102)$$

where the development of the turbulence effect on the source term $\overline{S}_i$ has not being carried out, for the sake of simplicity. The model based on Eqs. (3.88a, 3.102, 3.96, 3.98) and their ancillary definitions is called the **$k - \varepsilon$ turbulence model**.

Two parts of the $CS$ are usually critical in assigning BCs for Eqs. (3.96,3.98):

- at $CV$ inlets, it is important to specify correct or realistic values via first-kind conditions for $k$ and $\varepsilon$, because the inlet turbulence can significantly affect the downstream flow and consequently the other transport mechanisms also
- at solid surfaces, beside the no-slip condition that has to be satisfied, turbulence is also perturbed by the presence of the walls. Very close to the wall in the so-called "laminar sublayer," viscous damping reduces the tangential velocity fluctuations, while kinematic blocking reduces the normal fluctuations. Toward the outer part of the near-wall region, however, the turbulence is rapidly augmented by the production of turbulence kinetic energy due to the large gradients in mean velocity. Therefore, the treatment of this region significantly impacts the fidelity of turbulence modeling: these surfaces are the main source of mean vorticity and turbulence, and all variables of the governing equations have large gradients there, where momentum and other mechanisms are transported most vigorously. Traditionally, there are two approaches to model this near-wall region. In one approach, semiempirical formulas called "wall functions" (usually logarithmic) are used to bridge the laminar sublayer and the fully turbulent region. In another approach, a specially increased grid is adopted to allow for the governing Equations to be valid all the way to the surface, including the sublayer.

A common variation of this method is the $k - \omega$ model:[47] the first variable, $k$, determines the turbulence energy in the flow field, whereas the second variable (whether $\varepsilon$ or $\omega$) determines the scale of the turbulence (whether length scale or timescale). These two models have become industry standards and are commonly used for most types of engineering problems, especially in multiphysics frameworks.

---

[47] As proposed by Wilcox, D.C.: Turbulence Modeling for CFD. DCW Industries, La Cañada (1998).

## 3.4   Dimensionless Equations of Fluid Mechanics

The concepts presented in Sect. 2.4 (p. 49) can be worked further to identify the controlling dimensionless parameters that are commonly used in the applications of momentum transfer and related mechanisms. Let us invoke the governing Equations for laminar flow for a fluid with *constant rheological properties*, with *no source/sink term* and *in the steady state*

$$\frac{\partial u}{\partial x} + \frac{\partial v}{\partial y} = 0 \tag{3.57}$$

$$\rho_f \left( u \frac{\partial u}{\partial x} + v \frac{\partial u}{\partial y} \right) = \mu \left( \frac{\partial^2 u}{\partial x^2} + \frac{\partial^2 u}{\partial y^2} \right) - \frac{\partial p}{\partial x} \tag{3.69a revisited}$$

$$\rho_f \left( u \frac{\partial v}{\partial x} + v \frac{\partial v}{\partial y} \right) = \mu \left( \frac{\partial^2 v}{\partial x^2} + \frac{\partial^2 v}{\partial y^2} \right) - \frac{\partial p}{\partial y} \tag{3.69b revisited}$$

to solve for $\mathbf{w} = \mathbf{w}(x, y)$ and $p = p(x, y)$, for the $CV$ in Fig. 3.12. Then, let us choose some suitable dimensionless independent and dependent variables, relative to a flow having reference velocity and pressure of $\mathbf{w}_\infty$ and $(\rho_f \mathbf{w}_\infty)$, respectively; the flow interacts with a generic geometry having a reference length $L$. We can choose therefore:

$$\boxed{x^* \equiv \frac{x}{L}, \quad y^* \equiv \frac{y}{L}, \quad u^* \equiv \frac{u}{\mathbf{w}_\infty}, \quad v^* \equiv \frac{v}{\mathbf{w}_\infty}, \quad p^* \equiv \frac{p}{\rho_f \mathbf{w}_\infty^2}} \tag{3.103}$$

We proceed by operating the change of variable as performed already in Sect. 2.4.1 (p. 49), that is, by substituting variables of Eq. (3.103) in Eqs. (3.57) and (3.69) above, to obtain the **dimensionless Navier–Stokes Equations**:

$$\boxed{\begin{aligned} \frac{\partial u^*}{\partial x^*} + \frac{\partial v^*}{\partial y^*} &= 0 \\[2mm] u^* \frac{\partial u^*}{\partial x^*} + v^* \frac{\partial u^*}{\partial y} &= \frac{1}{Re} \left( \frac{\partial^2 u^*}{\partial x^{*2}} + \frac{\partial^2 u^*}{\partial y^{*2}} \right) - \frac{\partial p^*}{\partial x^*} \\[2mm] u^* \frac{\partial v^*}{\partial x^*} + v^* \frac{\partial v^*}{\partial y} &= \frac{1}{Re} \left( \frac{\partial^2 v^*}{\partial x^{*2}} + \frac{\partial^2 v^*}{\partial y^{*2}} \right) - \frac{\partial p^*}{\partial y^*} \end{aligned}}$$

$$\tag{3.104a}$$
$$\tag{3.104b}$$
$$\tag{3.104c}$$

The Reynolds number Re defined earlier

$$\mathrm{Re}_L \equiv \frac{\rho_\mathrm{f} \mathbf{w}_\infty L}{\mu} \qquad \text{(3.2 revisited)}$$

was purposefully employed. In the end, we see that the functions we sought were, **in dimensional form**:

$$\mathbf{w} = \mathbf{w}\,(x, y, L, \rho_\mathrm{f}, \mu, \mathbf{w}_\infty) \text{ and } p = p\,(x, y, L, \rho_\mathrm{f}, \mu, \mathbf{w}_\infty) \qquad (3.105)$$

while we have obtained, **in dimensionless form**:

$$\boxed{\mathbf{w}^* = \mathbf{w}^*\left(x^*, y^*, \mathrm{Re}_L\right) \text{ and } p^* = p^*\left(x^*, y^*, \mathrm{Re}_L\right)} \qquad (3.106)$$

with the resulting ease of discussion.

As for the appropriate BCs, the ones reported on earlier in Sect. 3.3.4 (p. 94) apply.

## 3.5 Momentum Transfer Interactions: The Boundary Layer

So far, we have described the physical meaning of velocity profiles in elementary flows as in Sect. 3.2 (p. 74). Even in those cases, the flows were affected by solid/fluid interaction due to limiting solid surfaces, but flow deformation is important in a variety of processes of our interest. The topic has been extensively reported by Schlichting [6], and only the main results are presented here.

The underlying concept to flow deformation in the laminar regime is the velocity boundary layer[48] that is important in a multiphysics framework as it is linked to the transport of any PCB quantity: the convective heat and mass transfer between a surface and a fluid, for example, all happen in the boundary layer region.[49] In contrast, *free-stream* flows are those in which the velocity is negligibly influenced by shear effects, that is, it is dominated by inertial and pressure effects only. These situations are also called inviscid flows, already encountered in Sect. 3.3.5 (p. 96).

---

[48] After the aforementioned physicist L. PRANDTL.

[49] Consequently, we will have a *thermal boundary layer* (treated in Chap. 4) and/or a *concentration boundary layer* (treated in Chap. 5). Their extension can also be slightly different than the velocity boundary layer, in the normal-to-wall or crosswise direction, as we will see in due time.

Whenever momentum transfer occurs, the flow deforms considerably with the confinement effect offered by plates or tubes start to occur (entrance effect or *developing flow*). Sufficiently downstream from disturbance onset, chances are that the flow will become uniform, i.e., its deformation is independent on both longitudinal coordinate and flow direction (*fully developed flow*). Moreover, a number of localized fluid patterns may result owing to geometry effects (such as *stagnation* or *flow separation*). It is evident then that we need to take a closer look to the velocity boundary layer phenomenology.

Velocity boundary layers are purposeful to the scale or OoM analysis, to simplify and assess equations with many terms. Velocity boundary layers can be also computed analytically, but frequent departures of geometry from ideality and boundary condition non-uniformity are found in the applications so that these methods involve mathematical difficulties. When faced to technical challenges, these difficulties are unjustified in the presence of proper numerical solutions and computing power.

### 3.5.1  Onset, Scale Analysis, Analytical Solutions

Let us consider the simplest external flow: a flow parallel to a fixed horizontal plate, very long in its direction. The fluid approaches the plate with an undisturbed (uniform) velocity profile $\mathbf{w}_\infty$ (Fig. 3.17, profile at left); as anticipated in Sect. 3.1.2 (p. 70), a deforming action due to fluid viscosity will start to propagate normally to the plate (Fig. 3.17, profile at right).

Then, along its course, the perturbation grows, deforming even more: a *developing velocity profile* (Fig. 3.18). At the steady state, the space neighboring the plate is characterized by a perturbation thickness that varies along distance or flow direction $x$. The flow field is then subdivided in the thin region next to the plate, in which the viscous effect must be taken into account (the *boundary layer* proper), and the contiguous region, where the viscous stresses are small and can be neglected (the *inviscid external region*). The boundary layer thickness, indicated by $\delta(x)$, is the distance over which the horizontal component $u(x, y)$ of velocity $\mathbf{w}$ changes from the undisturbed value $\mathbf{w}_\infty$ to 0 at the solid surface. Eventually, after a sufficiently long distance, $\delta$ will grow no longer and the boundary layer will be described by a *fully developed velocity profile*.

As anticipated in Sect. 3.1.1 (p. 67), a laminar flow regime occurs below a certain velocity magnitude threshold. With this geometry, the flow behavior can be assessed by a *local* Reynolds number depending on $x$, so its definition Eq. (3.2) is rewritten as

$$\mathrm{Re}_x \equiv \frac{\rho_\mathrm{f} \langle w \rangle x}{\mu} \tag{3.107}$$

and initial laminar flow is maintained until $x$ does not exceed a certain limit value.

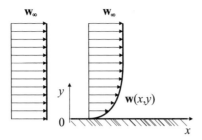

**Fig. 3.17** Flow parallel to a plate: the perturbation propagates in the otherwise undisturbed velocity distribution

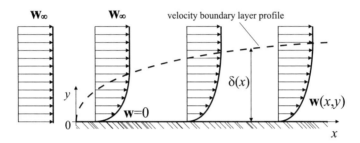

**Fig. 3.18** Flow parallel to a plate: the velocity distribution $\mathbf{w}(x, y)$ is being deformed along $x$, in the developing boundary layer, whose thickness $\delta$ varies along $x$

In order to illustrate the dependence of boundary layer features such as *thickness* and *skin friction* on the *flow regime*, it is appropriate to perform a scale analysis on governing Eqs. (3.57, 3.69) seen earlier in the usual assumption of *incompressible flow with constant rheological properties*, and with no mass force terms (which are negligible for this configuration, as with strong horizontal flow, and no thermal energy involved). Now, let us assume that thickness $\delta$ (Fig. 3.18) is always very small when compared to $x$:

$$\boxed{\delta \ll x}$$ (3.108)

In other words, we assume that **the boundary layer is thin**. Let us start with considering Eq. (3.69a); last inequality allows us to simplify it in two ways:

1. the scales of the first derivatives of $u$ are the following:

$$\frac{\partial u}{\partial x} \sim \frac{-\mathbf{w}_\infty}{x} \tag{3.109a}$$

$$\frac{\partial u}{\partial y} \sim \frac{\mathbf{w}_\infty}{\delta} \tag{3.109b}$$

As a consequence, the second derivatives can be evaluated as

$$\frac{\partial^2 u}{\partial x^2} \sim \frac{\frac{\partial u}{\partial x}\big|_{x=x} - \frac{\partial u}{\partial x}\big|_{x=0}}{x - 0} \sim \frac{\frac{-\mathbf{w}_\infty}{x} - 0}{x} \sim \frac{-\mathbf{w}_\infty}{x^2} \tag{3.110a}$$

$$\frac{\partial^2 u}{\partial y^2} \sim \frac{\frac{\partial u}{\partial y}\big|_{y=\delta} - \frac{\partial u}{\partial y}\big|_{y=0}}{\delta - 0} \sim \frac{0 - \frac{\mathbf{w}_\infty}{\delta}}{\delta} \sim \frac{-\mathbf{w}_\infty}{\delta^2} \tag{3.110b}$$

Comparing these two results, we come up with

$$\frac{\partial^2 u/\partial x^2}{\partial^2 u/\partial y^2} \sim \left(\frac{\delta}{x}\right)^2 \ll 1 \tag{3.111}$$

The thinness of the boundary layer implies that the $\partial^2 u/\partial x^2$ term can be neglected while considering Eq. (3.69a) inside the boundary layer. In other words, the *diffusive* gradients along the plate can be neglected with respect to those normal to the plate.

2. the pressure $p$ does not vary appreciably across the thin boundary layer:

$$p(x, y) \cong p(x) \tag{3.112}$$

Therefore, the pressure inside the boundary layer, $p(x)$, must be equal to the pressure at the outer edge of the boundary layer, $p_\infty$, which is constant:

$$\frac{\partial p}{\partial x} \cong \frac{dp}{dx} = \frac{dp_\infty}{dx} = 0 \tag{3.113}$$

The thinness of the boundary layer implies also that we can eliminate the unknown pressure from the problem; therefore, we can drop the unnecessary Eq. (3.69b).

After the deductions of Eqs. (3.111, 3.113), the problem is left with the continuity Eq. (3.57) and the Equation of momentum conservation along $x$ Eq. (3.69a), which reduces to the following *boundary layer momentum Equation*

$$u \frac{\partial u}{\partial x} + v \frac{\partial u}{\partial y} = \nu \frac{\partial^2 u}{\partial y^2} \qquad (3.114)$$

The two terms appearing in Eq. (3.114) hold the meaning of the *inertia or deceleration force* of the fluid $CV$, term ①, and *viscous or friction force* transmitted to the fluid $CV$ by the solid surface, term ②. Based on scale analysis similitudes of Eq. (3.109a), and the counterpart of Eq. (3.109b) written for the $v$ component of velocity, the continuity Eq. (3.57) can be rewritten as

$$\frac{\mathbf{w}_\infty}{x} \sim \frac{v}{\delta} \qquad (3.115)$$

while the Equation of momentum conservation along $x$ Eq. (3.114) is rewritten as

$$\mathbf{w}_\infty \frac{\mathbf{w}_\infty}{x} + v \frac{\mathbf{w}_\infty}{\delta} \sim \nu \frac{\mathbf{w}_\infty}{\delta^2} \qquad (3.116)$$

The two terms on the left-hand side have the same OoM (they can be also obtained by multiplying Eq. (3.117) by $\mathbf{w}_\infty$); therefore, Eq. (3.116) can be further simplified as

$$\mathbf{w}_\infty \frac{\mathbf{w}_\infty}{x} \sim \nu \frac{\mathbf{w}_\infty}{\delta^2} \qquad (3.117)$$

Equations (3.117, 3.115) deliver other two unknown scales:

$$\delta(x) \sim \sqrt{\frac{\nu x}{\mathbf{w}_\infty}} = x \mathrm{Re}_x^{-1/2} \qquad (3.118)$$

$$v \sim \sqrt{\frac{\nu \mathbf{w}_\infty}{x}} \qquad (3.119)$$

but it is Eq. (3.118) which draws our attention: *the boundary layer thickness depends directly on fluid kinematic viscosity and surface length and inversely on undisturbed fluid velocity*; that is, recalling the definition Eq. (3.107), **the thickness $\delta$ is directly proportional to surface length $x$ and inversely proportional to the square root of the Reynolds number** based on $x$, $\mathrm{Re}_x$.

The notion of local Re is at stake when **evaluating the scale of skin friction,** which is a form of *fluid resistance* or *drag*. Starting with the *local wall shear stress*:

$$\tau_{yx,x} = -\mu \left. \frac{\partial u}{\partial y} \right|_{y=0} \sim \mu \frac{\mathbf{w}_\infty}{\delta} \sim \mu \frac{\mathbf{w}_\infty}{x} \sqrt{\mathrm{Re}_x} \qquad (3.120)$$

Manipulating last term as

$$\sqrt{\text{Re}_x} = \sqrt{\text{Re}_x}\frac{\sqrt{\text{Re}_x}}{\sqrt{\text{Re}_x}} = \text{Re}_x\text{Re}_x^{-1/2} = \frac{x\mathbf{w}_\infty}{\nu}\text{Re}_x^{-1/2}$$

we get

$$\tau_{yx,x} \sim \frac{\rho_f\mathbf{w}_\infty^2}{\sqrt{\text{Re}_x}} \tag{3.121}$$

We commonly refer to this quantity, using the dimensionless form of the *local skin friction coefficient* $C_{fx}$,[50] as:

$$\boxed{C_{fx} \equiv \frac{\tau_{yx,x}}{\frac{1}{2}\rho_f\mathbf{w}_\infty^2}} \tag{3.122}$$

then it will be

$$\boxed{C_{fx} \sim \text{Re}_x^{-1/2}} \tag{3.123}$$

that is, **the greater the Reynolds number, the lesser the friction be offered by the solid surface**.

Two last implications of inequality Eq. (3.108):

1. combining Eq. (3.108) with Eq. (3.118), we see that the thin boundary layer assumption $\delta \ll x$ is verified when

$$\text{Re}_x^{1/2} \gg 1 \tag{3.124}$$

2. these boundary layer theory results hold for positions $x$ sufficiently far downstream from the leading edge of the surface, so that Eq. (3.124) is satisfied. So, the theory breaks down at $x = 0$ and for small $x$'s such that $\text{Re}_x^{1/2} \lesssim 1$.

The classical analytical solution of the boundary layer momentum Eq. (3.114), together with the continuity Eq. (3.57), can be based on the reduction of the PDE problem to an ordinary differential problem by using a *similarity transformation*.[51] Without digging in the demonstration [7], which is outside the scope of this book, we

---

[50]It is worth to note the similarities between local skin friction coefficient $C_{fx}$ and the Fanning friction factor previously defined by

$$f \equiv \frac{\tau_w}{\frac{1}{2}\rho_f\langle u\rangle^2} \tag{3.29}$$

With $C_{fx}$, the driving velocity is the undisturbed one, $\mathbf{w}_\infty$, and the resulting dimensionless group is $x$-dependent.

[51]As proposed by German physicist P.R.H. BLASIUS in the early twentieth century.

report the solution of the scales Eqs. (3.118, 3.123) for the boundary layer thickness $\delta(x)$ and local skin friction coefficient $C_{fx}$:

$$\delta(x) = 5.2x\mathrm{Re}_x^{-1/2} \tag{3.125}$$

$$C_{fx} = 0.664\mathrm{Re}_x^{-1/2} \tag{3.126}$$

### 3.5.2 Phenomenology

Let us give a closer look at the entrance/developing fluid dynamics involved in the two main configurations, by describing the resulting flow topology.

#### 3.5.2.1 Internal Flow

Let us consider first a confined flow in a tube. In this case, we will be examining the boundary layer progress along the tube length.

As the fluid enters the tube, the velocity at the wall $u(R, x)$ goes to 0, and due to mass continuity, the velocity profile becomes deformed, as velocity increases at centerline with respect the undisturbed value (compare the two velocity profiles in Fig. 3.19, bottom). It is evident, therefore, that for a confined flow the overall boundary layer is determined by two facing *developing* boundary layer regions (Fig. 3.19, mid), finally merging at the *entry length* $x'$. Actually, in full 3-D terms, the developing boundary layer is the surface obtained by rotating one single boundary layer by $2\pi$ radians about the tube centerline. An axial view of the boundary layer buildup is provided in Fig. 3.19, top, reporting on two velocity isosurfaces (with $u = $ constant).

If the flow is initially laminar, in the entry region of the flow (the core or *undisturbed zone*) the regime is still laminar: for increasing $x$s, the thickness of the boundary layer grows accordingly, until the tube is filled completely by the boundary layer as described. At this point, the entry length $x'$ divides the *developing flow* region from the *fully developed flow* region (Fig. 3.19, mid). It is worth noting that a *laminar sublayer* will always be implied, even in this configuration.

The fully developed flow will be laminar, if at $x'$ the flow is still such; otherwise, it will be turbulent. The turbulent profile, such as the one depicted in Fig. 3.1, right, is more uniform along $r$: although already developed, it is similar to the developing one, but it is maintained constant streamwise (along $x$). Due to mass continuity, the mass flow rate $\dot{m}$ is constant streamwise, and **the integral of the velocity profile along $r$ is always the same**, streamwise. Recalling the definition of mass flow rate

$$\dot{m} \equiv \rho_f \langle u \rangle \Omega \tag{3.16}$$

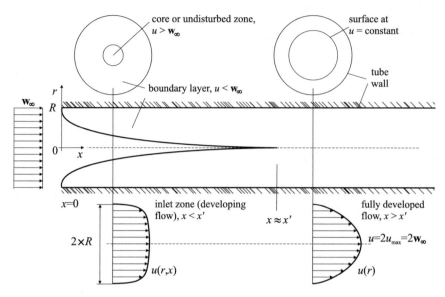

**Fig. 3.19** Flow zones and transition from laminar to turbulent boundary layer along a tube. Top: axial views of velocity isosurfaces at two different distances from inlet. Mid: transversal view of the boundary layer profile. Bottom: velocity distributions at two different distances from inlet

in such a configuration, the **volumetric flow rate** $\dot{V}$ may be given

$$\dot{V} \equiv \frac{\dot{m}}{\rho_{\mathrm{f}}} = \langle u \rangle \Omega \qquad (3.127)$$

Usually, the local Reynolds number for internal flow is given on the bulk velocity

$$\mathrm{Re}_x \equiv \frac{\langle w \rangle x}{\nu} \qquad (3.107 \text{ revisited})$$

In common situations, the *transition* from laminar to turbulent flow occurs for a critical value $\mathrm{Re}_x \approx \mathrm{Re}' = 2300$, but in well-controlled conditions a fully developed laminar flow can be achieved even for $\mathrm{Re}_x \approx 10^4$.

The entry length $x'$ can be usually evaluated, for a fully developed laminar flow, as $x_e \approx 0,03D \times \mathrm{Re}_x$: that is, for example, if $\mathrm{Re}_x = 2000$, then $x' = 60D$. For fully developed turbulent flow, $x'$ is usually smaller: $25D \leq x' \leq 50D$.

### 3.5.2.2    External Flow

Then, let us extend the conditions described in the preceding section, allowing an unrestrained (turbulent) flow parallel to the plate as in Fig. 3.20. In this case, we will be focussing on the boundary layer subregions as they form normal to the

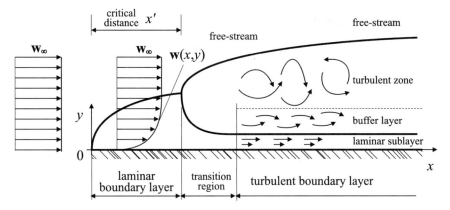

**Fig. 3.20** Flow nomenclature for the transition from laminar to turbulent boundary layer over a long plate, both streamwise and crosswise. A turbulent boundary layer is created through a transition from an initially laminar boundary layer

plate, during the onset of the developing flow. Obviously, these subregions appear, to various extensions, in any geometrical configurations.

Past the critical distance $x'$ from leading edge, the flow becomes unstable, as the local fluctuations of velocity do not die down but tend to augment instead. The region immediately past $x'$ is called the *transition region* from laminar to turbulent flow. The local fluctuations of the velocity **w** start to cause a macroscopic transport of momentum along $y$; then, many molecules are seen crossing the bulk flow, invigorating the fluid mixing. The value of $x'$ depends on many factors such as the shape of the leading edge, the turbulence of the free stream, and the roughness of the plate. In common situations, the transition occurs for a critical value $\mathrm{Re}_x \approx \mathrm{Re}' = 3.5 \times 10^5$. We had already some opportunity to pay attention at turbulent phenomenology when dealing with BCs of turbulent flows in Sect. 3.3.7.5 (p. 105). As illustrated on the right of Fig. 3.20, past the transition region, the flow can be subdivided crosswise in 4 regions:

1. a very thin *laminar* or *viscous sublayer* with prevailing molecular transport, in which the flow is laminar due to the strong **w** gradient (or $\partial u / \partial y$);
2. a *buffer layer*, featuring intense fluctuations of **w** but with a far smaller transversal transport (smaller $\partial u / \partial y$);
3. a wide *turbulent zone* where the fluctuations of **w** are less intense, but where there is a consistent transversal mixing ($\partial u / \partial y$ tends to 0);
4. a *free-stream* region where the flow is undisturbed by the plate.

With reference to the undisturbed velocity $\mathbf{w}_\infty$, we can then speak of **turbulent boundary layer for external flow**, referring to the above flow region ensemble.

### 3.5.3  Flow Visualization

Beside the simple geometries illustrated in Figs. 3.19 and 3.20, it is instructive to examine few other cases of external flow, which include the interaction with a solid surface. However, it should be kept in mind that the science of flow visualization is a vast study field, and more illustrative examples supplemented by the description of the various techniques employed can be found in the dedicated literature.[52]

First, in Fig. 3.21, the patterns produced past a cylinder facing an undisturbed flow are reported. These depend on the Reynolds number based on the undisturbed velocity and cylinder diameter:

$$\mathrm{Re}_D \equiv \frac{\mathbf{w}_\infty D}{\nu} \qquad\qquad (3.2 \text{ revisited})$$

Fig. 3.21 shows self-explanatory, as a variety of patterns are seen: vortex shedding, boundary layer separations, and transition from laminar to turbulent boundary layer. The repeating pattern of swirling vortices caused by the unsteady separation around blunt bodies is also called the "von Kármán street."[53] It is clear that the local skin friction coefficient $C_{fx}$ defined by Eq. (3.123) will vary in a complicated fashion around the cylinder in any flow regime.

The interactions between solid and fluid phases are as numerous as the possible configurations found in the applications. Flow around protrusions placed on long plates is common. Boundary layers are deformed with the flow past these 3-D objects, even in the presence of smoother shapes and moderate Reynolds numbers, as in Fig. 3.22.

The visualization techniques often involve tracing particles of various consistencies in these cases, to emphasize the stagnation and separation regions. Often, these objects are blunt: as in Fig. 3.23, with a flow past a rectangular, long protrusion. A variety of flow *loci* are found at the solid–fluid interphase: an upwind stagnation region, a separation region just above the protrusion's top, and the wake extending away along the bulk flow coordinate, well longer than a length corresponding to the protrusion's height.

Special configuration is possible to enhance the fluid/surface interaction, such as with jet impingement. Figure 3.24 presents a visualization case of a round impinging jet on a flat plate, for a moderate Reynolds number. A "train" of vortices is created at the jet-surrounding quiescent fluid that travel all the way to the target plate, laterally destroying the boundary layer produced by stagnating overpressure just underneath the jet, hence producing a strong increase in local skin friction.

---

[52] See for example:

- Van Dyke, M.: An Album of Fluid Motion. 76). The Parabolic Press, Stanford (1982)
- Color Plates and Flow Gallery. In: Smits, A.J., Lim, T.T. (Eds.) Flow Visualization: Techniques and Examples. Imperial College Press, London (2012).

[53] After the aforementioned engineer T. VON KÁRMÁN.

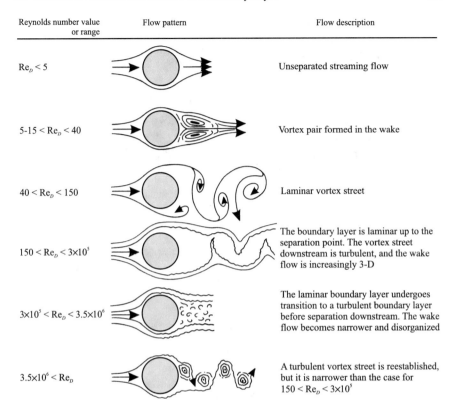

| Reynolds number value or range | Flow pattern | Flow description |
| --- | --- | --- |
| $Re_D < 5$ | | Unseparated streaming flow |
| $5\text{-}15 < Re_D < 40$ | | Vortex pair formed in the wake |
| $40 < Re_D < 150$ | | Laminar vortex street |
| $150 < Re_D < 3\times10^5$ | | The boundary layer is laminar up to the separation point. The vortex street downstream is turbulent, and the wake flow is increasingly 3-D |
| $3\times10^5 < Re_D < 3.5\times10^6$ | | The laminar boundary layer undergoes transition to a turbulent boundary layer before separation downstream. The wake flow becomes narrower and disorganized |
| $3.5\times10^6 < Re_D$ | | A turbulent vortex street is reestablished, but it is narrower than the case for $150 < Re_D < 3\times10^5$ |

**Fig. 3.21** Main flow regimes for flow across a cylinder. A variety of patterns are created *downwind* of the cylinder

**Fig. 3.22** Patterns of perturbed boundary layer due to the blocking action of a truncated cone

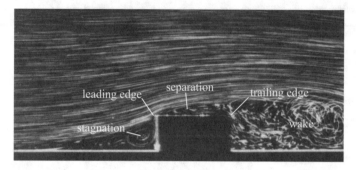

**Fig. 3.23** Main boundary layer *loci* for flow across a long rectangular cylinder. Patterns are created both *upwind* (before the *leading edge*) and *downwind* (after the *trailing edge*) of protrusion. So three characteristic regions are found: *stagnation*, *separation*, and *wake*

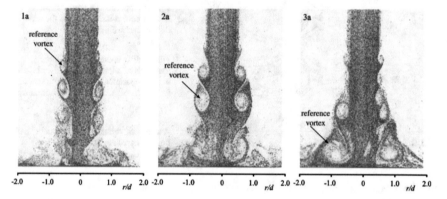

**Fig. 3.24** Particle Image Velocimetry (PIV) visualization of a round air impinging jet over a flat plate, with indication of a traveling reference vortex. Reprinted from International Journal of Heat and Mass Transfer, Vol 46, Michele Angioletti, Rocco M. Di Tommaso, Enrico Nino, Gianpaolo Ruocco, Simultaneous visualization of flow field and evaluation of local heat transfer by transitional impinging jets, 1703-1713, Copyright (2003), with permission from Elsevier

## 3.6  Numerical Solution of Fluid Flow

The driving Equations for fluid flow such as Eq. (3.70), or similar ones in other coordinate systems, must be solved analytically to determine the distribution of velocity, but they are solvable for a limited number of situations, only: these solutions are useful in helping to understand fluid flow but rarely can be used directly in engineering analysis, where more general and complex configurations occur.

We reported already in Chap. 1 and in Sect. 2.5 (p. 55) on the benefit of numerical solutions. In the case of fluid flow, the driving Equations are more complex to solve. To obtain an approximate solution, we will again use the FV method, treating the governing PDE for the transport of momentum in the same way as we did for energy. Moreover, the important supplementing concept of flow moving in the physical grid

**Fig. 3.25** $W$ and $E$ points surrounding the $P$, for the 1-D problem, and the $CV$ faces locations $w$ and $e$

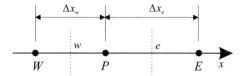

needs to be accommodated, and the essential ideas underlying the approach of FV for 3-D elliptic flows will be now presented. Following the same strategy laid out in Sect. 2.5, let us turn to a method to obtain the numerical (approximate) solution of these Equations, with a special look to the discretization approach and the iterative nature of the adopted method, as we have already determined which approach and data structure to exploit. The procedure will again allow for the transformation of the *differential* equation into a system of *algebraic* equations: to persuade the reader that numerical methods have rational motivations even for the solution of fluid flow problems. This treatment can again be carried forth by using the analogies among the heat and momentum transport, and foundations can be laid out for exploitation of somewhat automated computational procedures.

For the time being, only laminar flow will be examined and the treatment of turbulent flow is being postponed to the end of next chapter.

## 3.6.1 Discretization in 1-D

### 3.6.1.1 The New Terms

We will start with the geometry in Fig. 3.25 enforcing the FV again within the $CV$ along $x$, to identify a number of grid points and their interface points. For each point, an algebraic equation will be written containing the value of velocities for the cluster of points surrounding the subject grid point. Recalling the Navier–Stokes Equations, Eqs. (3.70) in their *variable properties and steady state* version:[54]

$$\frac{\mathrm{d}}{\mathrm{d}x}(\rho_f u) = 0 \tag{3.128a}$$

$$\underbrace{\frac{\mathrm{d}}{\mathrm{d}x}(\rho_f uu)}_{\text{①}} = \frac{\mathrm{d}}{\mathrm{d}x}\left(\mu \frac{\mathrm{d}u}{\mathrm{d}x}\right) \underbrace{- \frac{\mathrm{d}p}{\mathrm{d}x}}_{\text{②}} \pm S_x \tag{3.128b}$$

We now find two new items which were not present in the discrete equation for heat conduction such as Eq. (2.107): the **convective** term ① and the **gradient of pressure** term ②. We need to note that the discretization of the continuity Eq. (3.128a) also

---

[54]In order to comply with a non-Newtonian flow definition, the dynamic viscosity $\mu$ could well depend in a specific way on the local velocity gradient $\mathrm{d}u/\mathrm{d}x$.

requires attention. We also need to recognize that all of these new terms are *first derivatives in the space domain*.

### 3.6.1.2   The Convective Term: The Upwind Scheme

For the time being, let us leave out pressure term ② and the source term. Using Eq. (3.128a), the governing Eq. (3.128b) simplifies to

$$\rho_f u \frac{du}{dx} = \frac{d}{dx}\left(\mu \frac{du}{dx}\right) \tag{3.129}$$

Integrating Eq. (3.129) over the $CV$ of $P$ in Fig. 3.25, we get

$$(\rho_f uu)_e - (\rho_f uu)_w = \left(\mu \frac{du}{dx}\right)_e - \left(\mu \frac{du}{dx}\right)_w \tag{3.130}$$

To simplify the evaluation of the variable at the interface points, let us assume that $w$ and $e$ lay midway between neighboring grid points:[55]

$$u_e = \frac{1}{2}(u_E + u_P) \qquad \text{and} \qquad u_w = \frac{1}{2}(u_P + u_W) \tag{3.131}$$

then Eq. (3.130) can be written, using again the approximation of forward derivative

$$\frac{dT}{dx}\bigg|_{x_0} \approx \frac{T(x_0 + \Delta x) - T(x_0)}{\Delta x} \tag{2.106}$$

as

$$\frac{1}{2}(\rho_f u)_e (u_E + u_P) - \frac{1}{2}(\rho_f u)_w (u_P + u_W) = \mu_e \frac{u_E - u_P}{\Delta x_e} - \mu_w \frac{u_P - u_W}{\Delta x_w} \tag{3.132}$$

In order to employ the matrix representation already envisaged in Sect. 2.5 (p. 55), it is useful to cast Eq. (3.132) again in the following form:

$$a_P u_P + a_E u_E + a_W u_W = 0 \tag{3.133}$$

where

$$a_E = \frac{1}{2}(\rho_f u)_e - \frac{\mu_e}{\Delta x_e} \tag{3.134a}$$

---

[55]Recalling Note 46 in Chap. 2, this choice corresponds to the adoption of a *central difference scheme*.

$$a_W = -\frac{1}{2}(\rho_f u)_w - \frac{\mu_w}{\Delta x_w} \tag{3.134b}$$

$$a_P = \frac{1}{2}(\rho_f u)_e - \frac{1}{2}(\rho_f u)_w + \frac{\mu_e}{\Delta x_e} + \frac{\mu_w}{\Delta x_w}$$

$$= -(a_E + a_W) + (\rho_f u)_e - (\rho_f u)_w \tag{3.134c}$$

Equation (3.134c) has been easily worked out.[56] Moreover, due to mass continuity, it is $(\rho_f u)_e = (\rho_f u)_w$ and then Eq. (3.134c) simply becomes

$$a_P = -(a_E + a_W) \tag{3.135}$$

We soon realize that, when adding together a *specific flow rate* $\rho_f u$ and a *viscous diffusion term* $\mu/\Delta x$ with mixed signs, chances are that coefficients in Eqs. (3.134) do not comply with the important matrix property recalled in Note 46 in Chap. 2, ensuring physically sound results. To overcome this problem, an **upwind scheme** must be adopted.[57] Indeed, it is recognized here that the convected variable (such as the velocity $u$) at the interface points is numerically misleading (yielding non-smooth or checkerboard results) when computed as an average of neighboring grid point values. This convected variable, for example, $u_e$, should be assumed depending on the direction of the flow, instead; when employed the simplest of such schemes, it will then be

$$u_e = u_P \quad \text{if} \quad (\rho_f u)_e > 0 \tag{3.136a}$$

$$u_e = u_E \quad \text{if} \quad (\rho_f u)_e < 0 \tag{3.136b}$$

A similar definition can be indeed assumed for $u_w$. If so, this upwind scheme implies

$$(\rho_f u)_e \, u_e = \max\left[(\rho_f u)_e, 0\right] u_P - \max\left[-(\rho_f u)_e, 0\right] u_E$$

or, using a more compact formulation:

$$(\rho_f u)_e \, u_e = \|(\rho_f u)_e, 0\| \, u_P - \|-(\rho_f u)_e, 0\| \, u_E \tag{3.137}$$

For example, if the flow proceeds from left to right at point $e$ in the discrete geometry of Fig. 3.25, it is $(\rho_f u)_e > 0$ and Eq. (3.137) yields $u_e = u_P$. Using this position, Eqs. (3.134) become:

---

[56] This is simply done by adding and subtracting 1/2 of both velocity components and then reordering.

[57] Upwind schemes are a class of numerical discretization methods for solving hyperbolic PDEs. These schemes use an adaptive or solution-sensitive finite difference stencil to take into account the direction of propagation of information in a flow field. It is known that these schemes suffer somewhat from "numerical smear" or *false diffusion*, specially when the flow is not aligned with the grid; nevertheless, elaborated ways exist to alleviate the problem.

$$a_E = \|-(\rho_f u)_e, 0\| - \frac{\mu_e}{\Delta x_e} \tag{3.138a}$$

$$a_W = \|(\rho_f u)_w, 0\| - \frac{\mu_w}{\Delta x_w} \tag{3.138b}$$

$$a_P = -(a_E + a_W) \tag{3.138c}$$

### 3.6.1.3   Pressure-Gradient Term and Equation of Continuity: The Staggered Grid

Now, let us turn back to Eq. (3.128b). We soon realize that the pressure term ②, $dp/dx$, cannot be included in the computation of velocity just as a part of its *source/sink term*, for the pressure itself is an independent variable. It is possible to get rid of the pressure specification by means of vorticity-based methods [8] in 2-D geometries, however; but as we want emphasize the formal relationships between the variables that drive the transport phenomena, so the method outlined earlier in Sect. 2.5 will be again enforced.

The integration of $-dp/dx$ over the $CV$ of grid point $P$ in Fig. 3.25 yields the pressure drop $p_w - p_e$. To express this quantity in terms of grid point pressures, similarly than with Eqs. (3.131), we have

$$p_w - p_e = \frac{p_W + p_P}{2} - \frac{p_P + p_E}{2} = \frac{p_W + p_E}{2} \tag{3.139}$$

Then, we see that in the momentum Eq. (3.128b) for grid point $P$ the pressure difference between its two neighboring grid points will appear, with no information of the variable $p_P$ itself. Unfortunately, this is equivalent in using for $p$ a coarser grid than the one actually adopted, worsening the accuracy of the solution. Moreover, a somewhat alternating pressure field in all grid points along $x$, like the one in Fig. 3.26 (top), would imply a uniform but unrealistic pressure field, once its values are fit into the momentum Equation.

A similar difficulty occurs when integrating Eq. (3.128a) over the $CV$ of grid point $P$ in Fig. 3.25. We come up with $u_e - u_w = 0$. To express this quantity again in terms of grid point velocities, we have

**Fig. 3.26** Top: an alternating or wavy 1-D pressure field. Bottom: an alternating or wavy 1-D velocity field

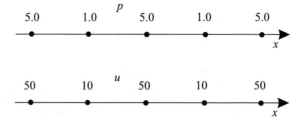

**Fig. 3.27** The calculation locations of $T$, $p$, and the staggered locations (white arrows) of $u$ for a 1-D grid

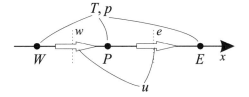

$$\frac{u_P + u_E}{2} - \frac{u_W + u_P}{2} = 0 \quad \text{or} \quad u_E - u_W = 0 \tag{3.140}$$

Again, we see that the discrete continuity Equation for grid point $P$ demands the equality of velocity at its two neighboring grid points, with the consequence that an unrealistic wavy velocity field along $x$, like the one in Fig. 3.26 (bottom), would satisfy continuity.

To overcome these problems, we may *calculate the variables on different locations just as well*: scalars such as temperature $T$ and pressure $p$ can still be calculated on the grid points such as $P$, while the velocity $u$ can be calculated on the $CV$ face locations such as $w$ and $e$. This grid choice, shown in Fig. 3.27, is called a **staggered grid**.

### 3.6.2 Discretization in 2-D

The upwind scheme and staggered grid ideas can be exploited when discretizing the momentum Equation in a multidimensional space. Let us consider the 2-D geometry in Fig. 3.28. As a consequence of the staggered grid adoption, the velocity component $u$ will be calculated on staggered locations (Fig. 3.29), just as well as for the other velocity component $v$ (Fig. 3.30). When required, all velocity values that are assigned on grid points $P$ can be evaluated based on the values occurring on $CV$ faces, by means of a linear interpolation; conversely, all property values that are assigned on $CV$ faces can be interpolated based on the values occurring at grid points.

Integrating the governing Equation in its unsteady state, variable property version

$$\rho_f \left[ \frac{\partial \mathbf{w}}{\partial \theta} + (\mathbf{w} \cdot \nabla) \mathbf{w} \right] = \mu \nabla^2 \mathbf{w} - \nabla p \pm \mathbf{S} \tag{3.70b}$$

for the velocity component $u$ in the geometry in Fig. 3.28, we have:

$$\int_{CV} \frac{\partial (\rho_f u)}{\partial \theta} dV + \int_{CV} \frac{\partial (\rho_f uu)}{\partial x} dV + \int_{CV} \frac{\partial (\rho_f vu)}{\partial x} dV =$$
$$\int_{CV} \frac{\partial}{\partial x} \left( \mu \frac{\partial u}{\partial x} \right) dV + \int_{CV} \frac{\partial}{\partial y} \left( \mu \frac{\partial u}{\partial y} \right) dV - \int_{CV} \frac{\partial p}{\partial x} dV + \int_{CV} S_x dV \tag{3.141}$$

**Fig. 3.28** A generic cluster of grid points ($W$, $E$, $N$, and $S$) surrounding the subject grid point $P$, for the 2-D problem. The dashed lines mark the $CS$ (faces at $w$, $e$, $n$, and $s$). $CV$ lengths are indicated

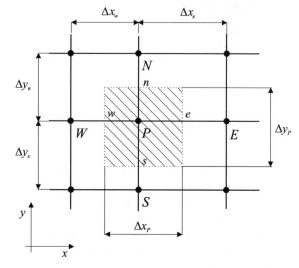

**Fig. 3.29** Scalars such as temperature and pressure are calculated at grid points (such as $P$, $N$, or $W$), whereas the velocity component $u$ is calculated at staggered locations

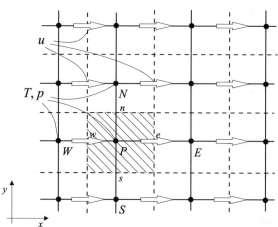

Based on the divergence theorem (Note 23), we can write:

$$\underbrace{\int_{CV} \frac{\partial\,(\rho_f u)}{\partial\theta}\,\mathrm{d}V}_{①} + \underbrace{\int_{CS} (\rho_f uu)\,\mathbf{i}\cdot\mathbf{n}\mathrm{d}S + \int_{CS} (\rho_f vu)\,\mathbf{j}\cdot\mathbf{n}\mathrm{d}S}_{②} =$$

$$\underbrace{\int_{CS} \left(\mu \frac{\partial u}{\partial x}\right)\mathbf{i}\cdot\mathbf{n}\mathrm{d}S + \int_{CS} \left(\mu \frac{\partial u}{\partial y}\right)\mathbf{j}\cdot\mathbf{n}\mathrm{d}S}_{③} - \underbrace{\int_{CV} \frac{\partial p}{\partial x}\mathrm{d}V}_{④} + \underbrace{\int_{CV} S_x \mathrm{d}V}_{⑤} \qquad (3.142)$$

with the usual notation of coordinates and normal surface versors. Now, let us develop each of these terms, from ① to ⑤.

**Fig. 3.30** The staggered locations for the calculation of velocity component $v$

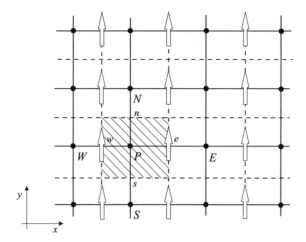

### 3.6.2.1 Unsteady Term ①

Using the fully implicit scheme for the discretization with time as in Sect. 2.5.3 (p. 61), we get

$$\int_{CV} \frac{\partial (\rho_f u)}{\partial \theta} dV = \frac{\rho_{fP} u_P - \rho_{fP}^0 u_P^0}{\Delta \theta} \Delta x_P \Delta y_P \tag{3.143}$$

### 3.6.2.2 Convective Terms ②

The two convective terms can be written by using the mass flow rates $F$ through $CS$:

$$\int_{CS} (\rho_f uu) \, \mathbf{i} \cdot \mathbf{n} dS + \int_{CS} (\rho_f vu) \, \mathbf{j} \cdot \mathbf{n} dS = F_e u_e - F_w u_w + F_n u_n - F_s u_s \tag{3.144}$$

with:

$$F_e = (\rho_f u)_e \Delta y_P \tag{3.145a}$$

$$F_w = (\rho_f u)_w \Delta y_P \tag{3.145b}$$

$$F_n = (\rho_f v)_n \Delta x_P \tag{3.145c}$$

$$F_s = (\rho_f v)_s \Delta x_P \tag{3.145d}$$

We then exploit the first-order upwind scheme (note that more elaborated and efficient schemes could be adopted) so that

$$F_e u_e = \|F_e, 0\| u_P - \|-F_e, 0\| u_E \tag{3.146a}$$

$$F_w u_w = \|F_w, 0\| u_W - \|-F_w, 0\| u_P \tag{3.146b}$$

$$F_n u_n = \|F_n, 0\| u_P - \|-F_n, 0\| u_N \tag{3.146c}$$

$$F_s u_s = \|F_s, 0\| u_S - \|-F_s, 0\| u_P \tag{3.146d}$$

### 3.6.2.3   Diffusive Terms ③

The two diffusive terms can be written by using the diffusive fluxes $D$ through $CS$:

$$\int_{CS} \left( \mu \frac{\partial u}{\partial x} \right) \mathbf{i} \cdot \mathbf{n} dS + \int_{CS} \left( \mu \frac{\partial u}{\partial y} \right) \mathbf{j} \cdot \mathbf{n} dS =$$
$$D_e(u_E - u_P) - D_w(u_P - u_W) + D_n(u_N - u_P) - D_s(u_P - u_S) \tag{3.147}$$

with:

$$D_e = \mu_e \Delta y_P / \Delta x_e \tag{3.148a}$$

$$D_w = \mu_w \Delta y_P / \Delta x_w \tag{3.148b}$$

$$D_n = \mu_n \Delta x_P / \Delta y_n \tag{3.148c}$$

$$D_s = \mu_s \Delta x_P / \Delta y_s \tag{3.148d}$$

### 3.6.2.4   Pressure-Gradient Term ④ and Source Term ⑤

For the pressure term, we can readily write:

$$-\int_{CV} \frac{\partial p}{\partial x} dV = (p_P - p_E)\Delta y_P \tag{3.149}$$

Just as readily, for the source/sink term we can write:

$$\int_{CV} S_x dV = S_x \Delta x_{CV} \Delta y_P \tag{3.150}$$

### 3.6.2.5   Rounding Up

Gathering Eqs. (3.143, 3.144, 3.147, 3.149, 3.150), the discrete form of Eq. (3.142) results:

$$\frac{\rho_{f P} u_P - \rho_{f P}^0 u_P^0}{\Delta \theta} \Delta x_P \Delta y_P +$$

$$\|F_e, 0\| u_P - \|-F_e, 0\| u_E - \|F_w, 0\| u_W + \|-F_w, 0\| u_P +$$

$$\|F_n, 0\| u_P - \|-F_n, 0\| u_N - \|F_s, 0\| u_S + \|-F_s, 0\| u_P =$$

$$D_e(u_E - u_P) - D_w(u_P - u_W) + D_n(u_N - u_P) - D_s(u_P - u_S) +$$

$$(p_P - p_E)\Delta y_P \pm S_x \Delta x_P \Delta y_P$$

$$(3.151)$$

Now, we can integrate the continuity Equation in its unsteady state, variable property version

$$\frac{\partial \rho_f}{\partial \theta} + \nabla \cdot (\rho_f \mathbf{w}) = 0 \tag{3.70a}$$

which is

$$\int_{CV} \frac{\partial \rho_f}{\partial \theta} dV + \int_{CV} \frac{\partial (\rho_f u)}{\partial x} dV + \int_{CV} \frac{\partial (\rho_f v)}{\partial y} dV = 0 \tag{3.152}$$

so we arrive at

$$\frac{\left(\rho_{f P} - \rho_{f P}^0\right) \Delta x_P \Delta y_P}{\Delta \theta} + F_e - F_w + F_n - F_s = 0 \tag{3.153}$$

using the definitions in Eqs. (3.145). The discrete form of the $x$-component of the Navier–Stokes Equation turns out to be:

$$\boxed{a_P u_P + a_E u_E + a_W u_W + a_N u_N + a_S u_S = (p_P - p_E)\Delta y_P + b_u} \tag{3.154}$$

where

$$a_E = -(D_e + \|-F_e, 0\|) \tag{3.155a}$$

$$a_W = -(D_w + \|F_w, 0\|) \tag{3.155b}$$

$$a_N = -(D_n + \|-F_n, 0\|) \tag{3.155c}$$

$$a_S = -(D_s + \|F_s, 0\|) \tag{3.155d}$$

$$a_P^0 = -\frac{\rho_{f P}^0 \Delta x_P \Delta y_P}{\Delta \theta} \tag{3.155e}$$

$$a_P = -(a_E + a_W + a_N + a_S + a_P^0) \tag{3.155f}$$

$$b_u = a_P^0 u_P^0 \pm S_x \Delta x_P \Delta y_P \tag{3.155g}$$

after some algebraic manipulation.[58] We cannot fail to note the similarity with Eq. (2.128) for the solution of heat conduction

$$a_P T_P + a_E T_E + a_W T_W + a_N T_N + a_S T_S = b \qquad (2.128)$$

Once spread over every elementary $CV$ of the system, Eq. (3.154) together with the definitions of Eqs. (3.145, 3.148, 3.155) gives out a coefficient matrix resembling again Eq. (2.115)

$$
\begin{bmatrix}
a_{P1} & a_{E1} & 0 & 0 & 0 & 0 \\
a_{W2} & a_{P2} & a_{E2} & 0 & 0 & 0 \\
0 & a_{W3} & a_{P3} & a_{E3} & 0 & 0 \\
\cdots & \cdots & \cdots & \cdots & \cdots & \cdots \\
0 & 0 & 0 & a_{W(n-1)} & a_{P(n-1)} & a_{E(n-1)} \\
0 & 0 & 0 & 0 & a_{Wn} & a_{Pn}
\end{bmatrix}
\begin{bmatrix}
T_1 \\ T_2 \\ T_3 \\ \cdots \\ T_{n-1} \\ T_n
\end{bmatrix}
=
\begin{bmatrix}
b_1 \\ b_2 \\ b_3 \\ \cdots \\ b_{n-1} \\ b_n
\end{bmatrix}
\qquad (2.115)
$$

with nonzero values again aligned along five diagonals. Note 46 of Chap. 2 also applies.

A similar development can be carried over for any remaining component of velocity.

---

[58] In order to enforce the conservative property dictated by the continuity Equation, and incorporate it into the discrete equation, we can multiply Eq. (3.153) by $u_P$ and subtract it from Eq. (3.131), getting

$$\frac{\rho_{fP}^0 \Delta x_P \Delta y_P}{\Delta \theta}(u_P - u_P^0) - (F_e - F_w + F_n - F_s)u_P +$$

$$\|F_e, 0\| u_P - \|-F_e, 0\| u_E - \|F_w, 0\| u_W + \|-F_w, 0\| u_P +$$

$$\|F_n, 0\| u_P - \|-F_n, 0\| u_N - \|F_s, 0\| u_S + \|-F_s, 0\| u_P =$$

$$D_e(u_E - u_P) - D_w(u_P - u_W) + D_n(u_N - u_P) - D_s(u_P - u_S) +$$

$$(p_P - p_E)\Delta y_P \pm S_x \Delta x_P \Delta y_P$$

Due to the continuity, the sum of mass flow rates $F$ with their sign is zero; grouping the coefficient, we have:

$$\frac{\rho_{fP}^0 \Delta x_P \Delta y_P}{\Delta \theta}(u_P - u_P^0) +$$

$$(D_e + \|F_e, 0\| + D_w + \|-F_w, 0\| + D_n + \|F_n, 0\| + D_s + \|-F_s, 0\|) u_P -$$

$$(D_e + \|-F_e, 0\|) u_E - (D_w + \|F_w, 0\|) u_W - (D_n + \|-F_n, 0\|) u_N - (D_s + \|F_s, 0\|) u_S =$$

$$(p_P - p_E)\Delta y_P \pm S_x \Delta x_P \Delta y_P.$$

### 3.6.3 The Solution of the Pressure Field

Within the FV, many levels of iterations are usually performed. This line of thought was introduced since the early 70s of last century by D.B. Spalding and speculated by S.V. Patankar [9] with the Semi-IMplicit Pressure Linked Equation (SIMPLE) method. This method with its many variants, that flourished during the last decades, is a set of pressure-based methods widely used for incompressible flow applications, which are the easiest ones to be attacked by a numerical solution. The primary idea behind SIMPLE is to create a discrete equation for pressure from the discrete continuity Equation. Since the continuity Equation contains discrete face velocities, SIMPLE provides a way, by means of the discrete momentum equations, to relate these discrete velocities to the discrete pressure field.

The method is based on two steps:

**The velocity-correction formula** It is clear that the system of Eq. (3.154) can be solved only when the pressure is given or is somehow estimated. Unless the correct pressure field is employed, the resulting velocity field will not satisfy the continuity Equation. With the SIMPLE method, this imperfect velocity field is based on an initial, guessed pressure field $\check{p}$; then, this field is purposely denoted by $\check{u}$, resulting from the solution of the following discrete formula:

$$a_P \check{u}_P + a_E \check{u}_E + a_W \check{u}_W + a_N \check{u}_N + a_S \check{u}_S = (\check{p}_P - \check{p}_E)\Delta y_P + b_u \qquad (3.156)$$

Therefore, we need to find a way of improving the guessed pressure $\check{p}$ such that the resulting velocity field $\check{u}$ will iteratively and progressively get closer to satisfying the continuity Equation. There are several ways to achieve this, but the simplest idea is to obtain the correct pressure $p$ from a *pressure correction* $\tilde{p}$:

$$p = \check{p} + \gamma \tilde{p} \qquad (3.157)$$

with $\gamma$ a suitable positive factor smaller than 1.[59] Correspondingly, the correction $(u', v')$ in the velocity field is determined in a similar way:

$$u = \check{u} + \tilde{u} \qquad (3.158)$$

Subtracting Eq. (3.156) from Eq. (3.154), we get

$$a_P \tilde{u}_P + a_E \tilde{u}_E + a_W \tilde{u}_W + a_N \tilde{u}_N + a_S \tilde{u}_S = (\tilde{p}_P - \tilde{p}_E)\Delta y_P \qquad (3.159)$$

One of the notable features of the SIMPLE method is that the sum of corrective terms neighboring the grid point $P$ Eq. (3.159), $(a_E \tilde{u}_E + a_W \tilde{u}_W + a_N \tilde{u}_N + a_S \tilde{u}_S)$, is taken as zero. We then end up with

---

[59]This *under-relaxation* coefficient depends on the specific SIMPLE variant. In the original version, $\gamma = 0.8$.

$$\tilde{u}_P = \frac{(\tilde{p}_P - \tilde{p}_E)\Delta y_P}{a_P} \tag{3.160}$$

referred to as the *velocity-correction formula*, as it can be plugged back into Eq. (3.158) to give

$$u_P = \check{u}_P + \frac{(\tilde{p}_P - \tilde{p}_E)\Delta y_P}{a_P} \tag{3.161}$$

**The pressure-correction Equation**    Next, we turn back to the continuity Equation. If we plug into Eq. (3.153), the velocity-correction formulas such as Eq. (3.160), we get[60]

$$a_{P\tilde{p}}\tilde{p}_P + a_{E\tilde{p}}\tilde{p}_E + a_{W\tilde{p}}\tilde{p}_W + a_{N\tilde{p}}\tilde{p}_N + a_{S\tilde{p}}\tilde{p}_S = b_{\tilde{p}} \tag{3.162}$$

where now

$$a_{E\tilde{p}} = -\frac{\rho_{fe}\Delta y_P^2}{a_e} \tag{3.163a}$$

$$a_{W\tilde{p}} = -\frac{\rho_{fw}\Delta y_P^2}{a_w} \tag{3.163b}$$

$$a_{N\tilde{p}} = -\frac{\rho_{fn}\Delta x_P^2}{a_n} \tag{3.163c}$$

$$a_{S\tilde{p}} = -\frac{\rho_{fs}\Delta x_P^2}{a_s} \tag{3.163d}$$

$$a_{P\tilde{p}} = -(a_E + a_W + a_N + a_S) \tag{3.163e}$$

$$b_{\tilde{p}} = \frac{\left(\rho_{fP} - \rho_{fP}^0\right)\Delta x_P\Delta y_P}{\Delta\theta} + \check{F}_e - \check{F}_w + \check{F}_n - \check{F}_s \tag{3.163f}$$

using again the definitions in Eqs. (3.145). In this development, we have assumed that the density $\rho_f$ does not depend directly on pressure.[61]

We see that the matrix data structure to solve the equation system we chose earlier is found once again.

From inspection of Eq. (3.163f), it is seen that $b_{\tilde{p}}$ is, in absolute value, the left-hand side of continuity Eq. (3.153), in terms of guessed velocity $\check{u}$. If corrective $b_{\tilde{p}}$ happens to be zero, it would mean that the guessed velocity $\check{u}$, in conjunction with the available value of $(\rho_{fP} - \rho_{fP}^0)$, satisfies continuity, and no pressure correction $\tilde{p}$ is needed. Then, term $b_{\tilde{p}}$ is the mass imbalance which the pressure correction, together with the associated velocity correction, must bring to zero.

A similar development can be carried over for any remaining component of velocity.

---

[60]The subscript $\tilde{p}$ has been introduced to avoid confusion with previous discrete coefficients.

[61]In case of highly compressible flows, Eq. (3.162) needs proper modifications [9].

### 3.6.4  A Sequence for the Solution of the Incompressible Navier–Stokes Equations

Now, we take a look to the operation sequence:

1. guess pressure field $\check{p}$
2. solve the momentum Equation Eq. (3.156) to obtain $\check{u}$, as well as the associated Equations for other guessed velocity components
3. evaluate term mass imbalance, or $b_{\check{p}}$ by Eq. (3.163f), to check continuity compliance, at every grid point, under a specified tolerance for achieved convergence
4. solve the pressure-correction Eq. (3.162) for $\tilde{p}$
5. calculate pressure $p$ from Eq. (3.157)
6. calculate $u$ from $\check{u}$ using Eq. (3.160), applying an under-relaxation with $\gamma = 0.5$ according to

$$\frac{a_P}{\gamma} T_P + a_E T_E + a_W T_W = b + \frac{1-\gamma}{\gamma} a_P T_P^* \tag{2.121}$$

   as well as with the other velocity components via their associated discretized Equations
7. treat the corrected pressure $p$ as a new guessed pressure $\check{p}$
8. return to Step 2

This sequence will be carried over for each of the velocity components in the flow field at stake.

### 3.6.5  Boundary Equations

While forming the final algebraic system, two general forms like Eqs. (3.154) and (3.162) can be cast for any internal grid point, that is, for any situation depicted in Fig. 3.29. According to the discussion leading to the fundamental Equation of heat transfer by conduction

$$\frac{1}{\alpha} \frac{\partial T}{\partial \theta} = \nabla^2 T \pm \frac{\dot{e}'''}{\lambda} \tag{2.22}$$

so far we have classified boundaries according to the variable or its derivative specified there (or a combination thereof). For fluid flow problems, we must further distinguish between two occurrences:

- for the discrete Navier–Stokes Equations Eq. (3.154), *flow boundaries* exist where **flow enters or leaves the domain**, or *geometric boundaries* exist where **flow adheres with or it is limited by the domain interface** (with Note 12 in mind)
- for the pressure-correction Eq. (3.162), *pressure boundaries* can be encountered, where **normal velocity** or **static pressure** is given

For example, while studying a process oven we might not include the boiler from where the hot air is drawn to the chamber, but then we must specify pressure, temperature, velocity, etc., of the flow as it leaves the boiler and enters the oven, while the geometric boundaries would be the external walls of the oven. Flow boundaries may further be classified as *inflow* and *outflow* boundaries.

### 3.6.5.1 Flow and Geometric Boundary Condition

Let us consider that the $CV$ of grid point $P$ is limited by an open inflow boundary corresponding to the $CS$ portion in $B$ (Fig. 3.31). Here, as we are given the inlet velocity $u_B$,[62] such that:

$$\int_{CS} (\rho_f uu) \, \mathbf{i} \cdot \mathbf{n} dS = -F_w u_w = -(\rho_f uu)_B \, \Delta y_P \qquad (3.164)$$

Therefore, Eq. (3.154) changes into

$$a_P u_P + a_E u_E + a_N u_N + a_S u_S = (p_B - p_P)\Delta y_P + b_u \qquad (3.165)$$

Some coefficient definitions, Eqs. (3.155b, 3.155f, 3.155g), modify as follows, respectively:

$$a_B = -\frac{\mu_B \Delta y_P}{\Delta x_B} \qquad (3.166a)$$

$$a_P = -(a_E + a_B + a_N + a_S + a_P^0) \qquad (3.166b)$$

$$b_u = a_P^0 u_P^0 \pm S_x \Delta x_P \Delta y_P + (\rho_f uu)_B \, \Delta y_P \qquad (3.166c)$$

while the others, Eqs. (3.155a, 3.155c–3.155e), still apply. No additional conceptual difficulties arise when the inflow velocity $u_B$ is not aligned with the grid, whereas the projections of the inflow velocity vector are employed.

Now, let us consider that the $CV$ of $P$ is limited by an open outflow boundary in $B$ (Fig. 3.32). Here, we do not require the value of $u$ to be specified, as it is determined by the flow field in the domain and convected to the boundary by the exiting flow.[63]

---

[62] Inflow boundaries should be placed at locations where we have sufficient data, either from another numerical simulation or from experimental observations.

[63] To gain insight on the convection-diffusion concept, a useful dimensionless number is the grid Peclet number $\mathrm{Pe} = \rho_f u \Delta x / \mu$ (after French physicist J.C.E. PÉCLET, in the early nineteenth century) which is the ratio of the advection rate by the rate of diffusion rate (in this case for the $u$ component of velocity). It is seen here that $\mathrm{Pe} = \infty$: the flow rate is high enough so that conditions downstream of our boundary do not affect the solution inside the domain.

**Fig. 3.31** A cluster of grid points adjacent to an inflow port, with indication of the velocity in $B$, for a flow field 2-D problem

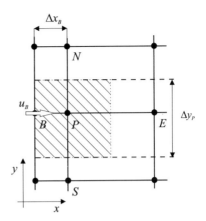

**Fig. 3.32** A cluster of grid points adjacent to an outflow port, with indication of the velocity in $B$, for a flow field 2-D problem

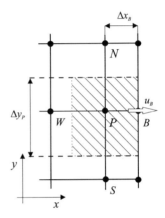

Therefore, it is $u_B = u_P$, so that there is no contribution in the flow leaving the domain due to diffusive flux:

$$\int_{CS} \left( \mu \frac{\partial u}{\partial x} \right) \mathbf{i} \cdot \mathbf{n} dS = D_e(u_B - u_P) = 0 \tag{3.167}$$

while the outlet flow is calculated as usual as

$$\int_{CS} (\rho_f uu) \, \mathbf{i} \cdot \mathbf{n} dS = F_e u_e = (\rho_f uu)_P \, \Delta y_P \tag{3.168}$$

Therefore, Eq. (3.154) changes into

$$a_P u_P + a_W u_W + a_N u_N + a_S u_S = (p_P - p_B)\Delta y_P + b_u \tag{3.169}$$

**Fig. 3.33** Recirculation at domain boundary, across section A

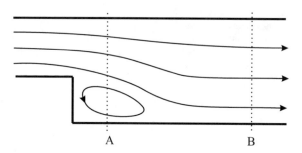

Some coefficient definitions, Eqs. (3.155a, 3.155f), modify as follows, respectively:

$$a_B = -(\rho_f u u)_P \, \Delta y_P \tag{3.170a}$$

$$a_P = -(a_B + a_W + a_N + a_S + a_P^0) \tag{3.170b}$$

while the others, Eqs. (3.155b-3.155e, 3.155g), still apply.

We must duly note that in this case the flow is directed out of the domain at all points on the $B$ boundary in Fig. 3.32. The choice of the appropriate location of such boundary must be made with care: the conditions downstream need to have no influence on the upstream solution. In fact, consider the flow past a backward-facing step, as shown in Fig. 3.33. If we choose location A as the outflow boundary, we cut across a recirculation pattern or *backflow*. In this situation, we would have to specify the $u$ (and $v$) values associated with the incoming portions of the flow for the problem to be well-posed. Location B, located well past the recirculation zone, is a much better choice for an outflow boundary.

Finally, in case that the $CV$ is limited by a solid–surface interface boundary in B, we can extend easily the notations of Eqs. (3.167, 3.168) by prescribing the value of $u_e$ and $u_B$: for example, in case of the common *no-slip* boundary condition, it will be simply $u_e = u_B = 0$ and the coefficients of Eq. (3.169) will be modified accordingly.

### 3.6.5.2 Pressure Boundary Conditions

The first case is when we are given the normal component of the velocity $\mathbf{w}_B$ (Fig. 3.34). This type of boundary could involve inflow or outflow boundaries, or a boundary normal to which there is no flow, such a walls. Equation (3.162) would read

$$a_{P\tilde{p}} \tilde{p}_P + a_{E\tilde{p}} \tilde{p}_E + a_{B\tilde{p}} \tilde{p}_B + a_{N\tilde{p}} \tilde{p}_N + a_{S\tilde{p}} \tilde{p}_S = b_{\tilde{p}} \tag{3.171}$$

but the flow rate $F_B$ can be expressed directly in terms of $u_B$, not in terms of $\breve{u}_B$ and a corresponding correction $\tilde{u}_B$. Therefore, no information of $\tilde{p}_B$ is needed, and $a_{B\tilde{p}}$ can be set as zero.

**Fig. 3.34** Near-boundary
$CV$ for pressure, with
indication of the velocity in
$B$

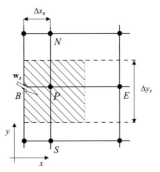

The second case is when the static pressure itself $p_B$ is specified at boundary. Equation (3.157) in this case will read $\check{p} = p_B$, and then, $\tilde{p}_B$ will be zero and Eq. (3.162) can be solved as a heat conduction equation with a boundary condition of the first kind.

### 3.6.5.3   The Relative Nature of Pressure

For incompressible flows, where density is not a function of pressure, the solution is usually acquired with more ease when flow boundaries are considered. In these cases, the level of pressure in the domain is not set: differences in pressure are unique, but the absolute pressure values themselves are not. This means that $p$ and $p + constant$ are solutions of the same problem.

From the computational viewpoint, we can see that the satisfaction of the continuity Equation represents an additional bond linking the mass balance in $N$ $CV$s, so that only $N - 1$ unique equations result. Therefore, we do not have enough equations for $N$ pressure (or pressure-correction) unknowns. This situation may be remedied by setting the pressure arbitrarily or alternatively $\tilde{p} = 0$ at one grid point in the domain. When the static pressure $p$ is given on a boundary, the pressure is made unique, and the problem does not arise.

For compressible flows, where the density is a function of pressure, it is necessary to specify pressure boundary conditions on at least one part of the domain boundary.

## 3.6.6   CFD Framework and Components

So far we implied that flows cannot be solved analytically except in special cases. To obtain an approximate solution numerically, we have to use a discretization method such as the one explained in Sects. 2.5.4, 3.6.2-3.6.4 (p. 63, p. 125–133, respectively) to form a huge system of algebraic equations, which can then be solved by a computer. Much as the accuracy of experimental data depends on the quality of the tools and signal treatment used, the accuracy of a numerical solution depends on the quality of discretization used. A common approach to accurate solving of complex transport phenomena problems is by means of the Computational Fluid Dynamics (CFD), through a qualified ensemble of software, hardware, and human resources. CFD is, in short, **a suite of data management and solution methods that can be used to solve any combination of transport phenomena PDEs**, even with a certain degree of automation, when proper engineering training is ensured.

Nowadays, in relation to multiphysics, the subject has broader scope than its name implies. Through the CFD, the engineer is now able to automate well-established engineering design methods, such as:

- prediction of the performance of new designs or processes before they are ever manufactured or implemented
- analysis of the flow and performance of preexisting process equipment in order to reach conclusions on the perspective improvement or optimization

The use of detailed numerical solutions of the governing PDEs as substitutes for experimental research into the nature of complex situations has become a common practice [10]. One can directly perform specific in-house programming or purchase design packages to solve simple problems in a few seconds on a personal computer, or may deal with integrated tasks involving hundreds of computing hours narrowing down to the slightest product/process/system detail. The combination of a calculation framework for complex flows and a satisfactory vehicle of heuristic but feasible methods for topics like turbulence and multiphase flow makes CFD a practical tool for industrial and environmental problems. Finally, as we try to convey in this book, CFD codes are perfectly suited for implementation up to full interaction among the PCB disciplines for the above multiscale ranges.

However, it should be insisted here that the advantages of CFD are conditional on the solution accuracy, bearing in mind that numerical results are, due to their nature, always approximate. Generally, departure of solution quality, with respect to "true" experimental values, arises due to the following:

- the governing PDEs may contain approximation or idealization;
- approximation is made in the discretization process;
- in solving the discretized equations, iterative methods are used, as implied earlier in Sect. 2.5.2.4 (p. 60), and the exact solution may be always inhibited even for repeated attempts.

In brief, a CFD suite allows the engineer to perform the following tasks:

1. parameters definition, including $CV$ geometry and (solid and fluid) media characteristics;
2. operating variables definition, such as temperature, velocity, pressure in the $CV$ and at its boundaries;
3. $CV$ geometry creation/import and manipulation;
4. PDE discretization and solution, by using a choice of numerical techniques;
5. collection of solution data and their organization to proper analysis and presentation.

Besides the FV discretization method mentioned earlier, other popular options exist. Moreover, as the CFD is mostly employed in the presence of momentum transfer, some special considerations on the solving grid need to be reported.

### 3.6.6.1 Numerical Discretization

Following the same considerations made already in Sect. 2.5.5 (p. 64), the *gridding strategy* should deformed to increase the grid points in the zones where greatest velocity gradients are expected. The solving system can then be solved by a suitable numerical algorithm.[64] Much as the accuracy of PCB experimental data depends on the quality of the tools and procedures employed, the accuracy of numerical solutions is dependent upon the quality of discretization used.

Another popular discretization procedure employs the *calculus of variations*, leading to the **Finite Element method** (FE). FEs are similar to FVs in many ways. The domain is broken into finite elements, generally unstructured (see below). But equations are multiplied by a weight function before they are integrated over the entire domain. In its simplest version, the FE solution is approximated by a linear shape function within each element in a way that guarantees continuity of the solution across element boundaries. Such a function can be constructed from its values at the corners of the elements. This approximation is then substituted into the weighted integral of the conservation equations, which are solved by requiring the derivative of the integral with respect to each nodal value to be zero; this corresponds to selecting the best solution within the set of allowed functions (the one with minimum residual). The result is a set of nonlinear algebraic equations. An important advantage of FE is the ability to deal with arbitrary geometries; the grid is easily refined, and each element can be easily subdivided. The drawback is that the solving matrix is not well structured, in the sense already implied in Sect. 2.5.2.4 (p. 60); therefore, the numerical system is less robust and can be prone to solution instabilities.

### 3.6.6.2 Geometrical Discretization

We have already seen in Sect. 2.5.5 (p. 64) some representations of *structured grids*. The way the $CV$ at stake is discretized is very much related to the choice of the numerical discretization scheme adopted. Grids may be constructed using a variety of cell shapes (Fig. 3.35). The most widely used are quadrilaterals (in 2-D) and hexahedra (in 3-D).

As CFD is used for analyzing industrial flows, *unstructured grids* are becoming necessary to handle complex geometries and boundary layers: it is shown in Fig. 3.36 that each grid point is connected to an arbitrary number of boundary points. Fig. 3.37

---

[64]It should be clear that the formal integration of differential equations by a computer program, although made possible by specialized software, is much more complicated than solving a system of algebraic equations.

**Fig. 3.35** Cell shapes:
triangle①, tetrahedron ②,
quadrilateral ③, hexahedron
④, prism ⑤, and pyramid ⑥

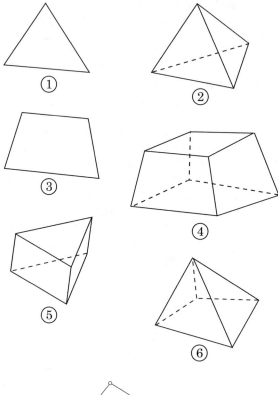

**Fig. 3.36** An unstructured
grid

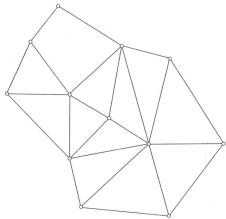

shows a *block-structured grid* instead. Here, the mesh is divided into blocks, and
the mesh within each block can be structured. However, the arrangement of the
blocks themselves is not necessarily structured. Grid blocks identify coarser and
finer grid levels; usually, on the former level, the blocks correspond to relatively large
segments of the domain; on the latter, a structured grid can be cast. Another possible

**Fig. 3.37** A
block-structured grid

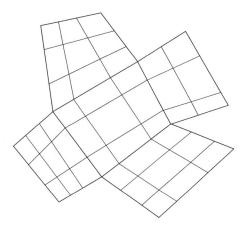

**Fig. 3.38** A hybrid grid
used for boundary layers

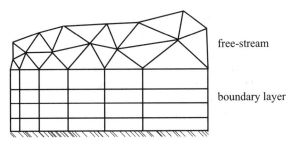

free-stream

boundary layer

arrangement is the hybrid grid used for boundary layers. For example, hexahedra are used within the boundary, transitioning to tetrahedra in the free stream (Fig. 3.38).

Nowadays, body-fitting gridding can be employed, and automatic generation is generally available within CFD suites, so that the grid aspect ratio can be controlled and the cells may be easily refined locally. Though mesh structure imposes restrictions, structured quadrilaterals and hexahedra are well-suited for flows with a dominant direction, such as boundary layer flows.

The flexibility is offset, though, by the disadvantage of the irregularity of the data structure: the solving matrix no longer has regular, diagonal structure; its bandwidth needs reduction by reordering of the grid points. This implies that the solver is usually slower than with regular grids, or even be impeded to converge. Overall, in case of very complex geometries, unstructured meshes impose fewer topological restrictions on the user and make it easier to mesh.

### 3.6.6.3 Properties of Solution Methods

As the numerical and geometrical discretization methods of choice lead to huge systems of nonlinear algebraic equations, direct or iterative solvers as implied in Sect. 2.5.4 (p. 63) can be selected and automatically applied to yield for the solution of the problem. It is important to define the main properties of the CFD solution, as briefly summarized in the following.

**Consistency**    The discretization process should yield a solving system whose numerical solution is equivalent to the analytical one, as the space (grid) and time discretizations tend to zero. The difference between the discretized equation and the analytical one is called the *truncation error* (already seen in Note 21 of Chap. 2), becoming zero in the limit of the discretization going to zero. Ideally, all PDE terms should be discretized with approximations of the same order of accuracy; however, some terms (e.g., convective terms in high Reynolds number flows or diffusive terms in low Reynolds number flows) may be dominant in a particular flow, and it may be reasonable to treat them with more accuracy than the others.

**Stability**    A solution method should not magnify the errors that appear in the course of solution process. For unsteady problems, stability guarantees that the method produces a bounded solution whenever the solution of the analytical equation is supposed to be bounded. For iterative methods, a stable method would not diverge. Stability can be difficult to investigate, especially when boundary conditions and nonlinearities are present. Many solution schemes require that the time step be smaller than a certain limit or that under-relaxation be used, as shown in Sect. 2.5.2.4 (p. 60).

**Convergence**    The final unchanging state of a numerical solution converges to the analytical one as the space and time discretizations tend to zero. For well-posed linear initial value problems, discretized by a finite difference approximation, that satisfy the consistency condition, stability is the necessary and sufficient condition for convergence.[65] For nonlinear problems which are strongly influenced by boundary conditions, stability and convergence are difficult to demonstrate; therefore, convergence can be checked using numerical experiments, i.e., by repeated calculations on a series of successively refined grids. If the method is stable and if all approximations used in the discretization process are consistent, we may find that the solution does converge to a *grid-independent* solution.

**Accuracy**    CFD solutions are always burdened with two other kinds of systematic errors: modeling errors, which are defined as the difference between the actual (experimental) situation and the analytical solution of the mathematical model, and iteration errors, defined as the difference between the iterative and exact solutions of the algebraic equations systems.

Errors from various sources may cancel each other, so that sometimes a solution obtained on a coarse grid may agree better with the associated experiment than a solution on a finer grid which, by definition, should be more accurate.

Modeling errors are the most critical in the analyst practice, as they depend on the assumptions made in deriving the governing PDEs. When dealing with laminar

---

[65] As stated by the *Lax equivalence theorem*, proposed by the Hungarian-born American mathematician P.L. LAX in mid-twentieth century. A solution whose behavior changes continuously with the initial conditions, if unique, is "well-posed." Inverse problems, deducing previous distributions of independent variable from final data, are often "ill-posed."

flows, they may be considered negligible when the Navier–Stokes Equations are invoked, as they represent a sufficiently accurate model of that flow; however, for turbulent flows, two-phase flows, reacting flow, and so on, modeling errors may be quite large, resulting in qualitatively wrong analytical solution.

Modeling errors are also introduced by simplifying the geometry of $CV$ and the boundary conditions, as well as other simplifying assumptions. These errors are not known a priori; they can only be evaluated by comparing solutions in which the discretization and convergence errors are negligible with accurate experimental data or with data obtained by more accurate models (e.g., data from direct simulation of turbulence, etc.).

## 3.7   Further Reading

- Fundamental laws; Fluid-flow phenomena; Turbulence. Potter, M.C., Foss, J.F.: Fluid Mechanics. Chapters 1–3, 5,7. Great Lakes Press, East Lansing (1982)
- Kinematics; Flow computation; Forces and stresses; Hydrostatics; Equation of motion and vorticity transport. Pozrikidis, C.: Fluid dynamics: theory, computation, and numerical simulation. Chapters 1–6. Springer, New York (2009)
- Heat and mass transfer in porous media, in forced and natural convection. Nield, D.A., Bejan, A.: Convection in Porous Media. Chapters 4–7. Springer, New York (2006)
- Dimensionless numbers in fluid dynamics and their meaning. Kuneš, J.: Dimensionless Physical Quantities in Science and Engineering. Elsevier, London (2012)
- Finite Volumes and development of SIMPLE variations; Elementary turbulence modeling. Versteeg, H.K., Malalasekera, W.: An Introduction to Computational Fluid Dynamics: the Finite Volume Method. Chapters 3–9. Longman, London (1996)
- Intro to Finite Elements concepts. Zienkiewicz, O.C.: The Finite Element Method. Chapter 3. McGraw-Hill, London (1977)
- Practical development of CFD codes, including turbulence. Nakayama, A.: PC-Aided Numerical Heat Transfer and Convective Flow. CRC press, Boca Raton (1995)
- Basic discretization techniques; The analysis of numerical schemes. Hirsch, C.: Numerical Computation of Internal and External Flows - Vol. 1. Chapters 4–8. John Wiley & Sons, New York (1988)
- Parabolic, elliptic, hyperbolic equations and related numerical methods. Ames, W.F.: Numerical Methods for Partial Differential Equations. Chapters 2–4. Academic press, New York (2014)
- Piping systems and design. Janna, W.S.: Design of Fluid Thermal Systems. Chapters 3–5. PWS Publishing, Boston (1998)
- The standard $k - \varepsilon$ turbulence model. Launder, B.E., Spalding, D.B.: Lectures in mathematical models of turbulence. Academic Press, London (1972)

# References

1. Liley, P.E., Thomson, G.H., Friend, D.G., Daubert, T.E., Buck, E.: Physical and chemical data. In: Green, D.W., J.O. Maloney (eds.) Perry's Chemical Engineers' Handbook, Chapter 2 (1997)
2. Irvine Jr., T.F.: Thermophysical properties. In: Rohsenow, W.M., Hartnett, J.P., Cho, Y.I. (eds.) Handbook of Heat Transfer. McGraw-Hill, New York (1998)
3. Bird, R.B., Stewart, W.E., Lightfoot, E.N.: Transport Phenomena. Wiley, New York (2002)
4. Bejan, A.: Heat Transfer. Wiley, New York (1993)
5. Griebel, M., Dornseifer, T., Neunhoeffer, T.: Numerical Simulation in Fluid Dynamics: a Practical Introduction. Society for Industrial and Applied Mathematics, Philadelphia (1998)
6. Schlichting, H.: Boundary-Layer Theory. McGraw-Hill, New York (1979)
7. Mauri, R.: Transport Phenomena in Multiphase Flows. Springer, New York (2015)
8. Anderson, D.A., Tannehill, J.C., Pletcher, R.H.: Computational Fluid Dynamics and Heat Transfer. Hemisphere Publishing, Washington (1984)
9. Patankar, S.V.: Numerical Heat Transfer and Fluid Flow. Hemisphere Publishing, Washington (1980)
10. Ferziger, J.H., Perić, M.: Computational Methods for Fluid Dynamics. Springer, Berlin (2002)

# Chapter 4
# Heat Transfer by Convection

**Abstract** Being the second mode that traditionally is encountered in the study of heat transfer, with the analysis of *convection* we learn on the heat transfer process that is *executed by a fluid stream*. The stream purposefully acts as a **heat carrier between two moving media in relative motion** coming in contact. It is clear, then, that this topic is focussed on the effects of *solid/fluid heat interaction*. So, while studying convection, we exploit the notions learned with the fluid mechanics that focussed on the effects of solid/fluid *dynamic* interaction. As we recall that convective heat flux is ruled by the description of a *heat transfer coefficient*, we describe briefly some basic devices where convection is realized following the *microscopic balance* leading to join the **distributions of the flow velocity vector and the temperature scalar**. It is clear that, at this point, we should distinguish between the temperature of the solid $T_s$, the one studied in Chap. 2, and the temperature of the fluid $T_f$ that will be the subject of this chapter, but whichever temperature we consider, we really face with the same scalar variable. However, for the sake of clarity sometime we indulge in differentiating the nomenclature, but unless we will be engaged in two-phase (solid/ fluid) modeling, we will hold that the subject at stake now is the fluid temperature $T_f$. Along the same lines that we exploit so far, then we derive and integrate the *governing differential equations* in various cases, and the concept of the *thermal boundary layer* is presented that carries analogies with the former fluid boundary layer. Finally, a *numerical solution* of the governing equations is completed, for the distribution of temperature and velocity, following the course cast so far.

## 4.1 Convection: The Underlying Physics and Basic Definitions

### 4.1.1 Macroscopical Heat Flow

Let us consider a certain fluid being heated from a hot plate below. In order to describe the onset of convection and related temperature distribution in the fluid, a *thermogram* can be employed as in Fig. 4.1. The heat released by the bottom plate

---

The original version of this chapter was revised: Belated corrections have been incorporated. The correction to this chapter is available at https://doi.org/10.1007/978-3-319-66822-2_7

G. Ruocco, *Introduction to Transport Phenomena Modeling*,
https://doi.org/10.1007/978-3-319-66822-2_4

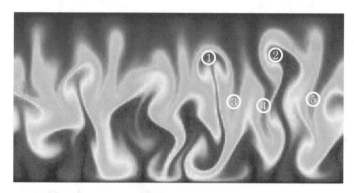

**Fig. 4.1** Warmer buoyant fluid structures ① ② and their cooler counterparts ③ ④ ⑤ in a natural convection configuration, as reported by thermal imaging

is carried along by the fluid, so that warmer plume structures are formed ① ② (red). The warmer fluid is readily replaced by cooler currents ③ ④ ⑤ (dark blue, white).

Thermal convection consists in the heat transfer realized by the flow of a fluid, in thermal contact with a solid surface being in thermal disequilibrium with it. Differently than with conduction, the subject medium is now in motion, with the characteristics described in the previous chapter: *The heat is macroscopically convected in the fluid flow, across and along the geometry at stake*. No additional microscopic mechanisms need to be invoked now, as the implied heat transfer mechanism between the surface and the fluid in the *laminar sublayer* seen in Sect. 3.5.2 (p. 115) is no other than pure conduction. It is clear then that the flow characteristics (such as velocity onset, driving force, boundary layer, turbulence) will greatly affect the study of this transport phenomena.

In addition to macroscopic transport of heat, we will learn that *any other scalar variable*, representing the descriptor of given PCB process of interest, *can be macroscopically transported by convection*: this is important specially for those issues that need to be invoked when dealing with turbulent flows.

As we already laid out all media properties that govern the workings of conduction and momentum transfer so far, we will deal now with some definitions that attain specifically heat transfer by convection.

### 4.1.2 Newton's Law of Convection

We will see that, as a result of the flow, velocity and temperature profiles are created in the fluid, obeying to geometry- and regime-specific *fluid dynamic boundary layer* and *thermal boundary layer*, respectively. For the time being, in this descriptive examples, no implications will be made whether the configuration is such that the

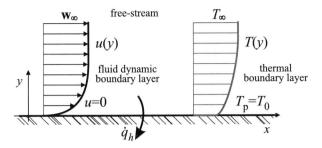

**Fig. 4.2** Qualitative velocity and temperature profiles in a fully developed flow along a *heated parallel plate*, with indication of thermal flux $\dot{q}_h$ exchanged between plate and flow: the flux is released by the fluid $CV$

limiting and perturbing surface is maintained at *uniform temperature*, or it is applied a *uniform heat flux* (some variation of this two main conditions are found in the applications).

### 4.1.2.1 Velocity and Temperature Profiles in an External Flow

Let us first consider a laminar flow of a fluid, whose motion is due to an external mechanical source, parallel to a plate in the steady state as in Fig. 4.2. Let the flow be in fully developed regime already, featuring an *undisturbed velocity* value $\mathbf{w}_\infty$, so that the velocity component $u$ depends solely on $y$. The flow is very wide along the other two coordinates: this large width (area), in any $x - z$ plane, will have a size $A \equiv \Delta x \times \Delta z$. The fluid's velocity is subject to the transition from the value at free stream $u = \mathbf{w}_\infty$, sufficiently far from the plate, to the no-slip condition $u = 0$ at the plate.

The temperature free stream is $T_\infty$. This exposed surface is set and maintained at a *uniform temperature* $T_{\mathrm{p}} = T_0$, with $T_{\mathrm{p}} < T_\infty$. Correspondingly, the fluid cools off from $T_\infty$ to $T_{\mathrm{p}}$: the heat lost by the fluid is represented by the uniform convective *thermal flux* $\dot{q}_h$, transferred to the plate and convected along with the fluid. In other words, the boundary conditions are

$$u(x, \infty) = \mathbf{w}_\infty, \quad u(x, 0) = 0, \quad T(x, \infty) = T_\infty, \quad T(x, 0) = T_0 \qquad (4.1)$$

### 4.1.2.2 Velocity and Temperature Profiles in an Internal Flow

Let us consider then a laminar flow of a fluid, whose motion is again due to an external source, in a tube in the steady state as in Fig. 4.3. Let the flow be in fully developed regime already, featuring a *nominal velocity* value $\mathbf{w}_\infty$, so that the velocity

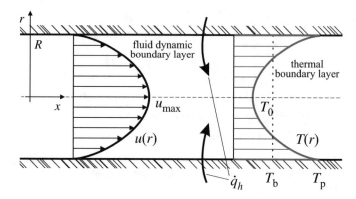

**Fig. 4.3** Qualitative velocity and temperature profiles in a fully developed flow flowing in a *cooled tube*, with indication of thermal flux $\dot{q}_h$ exchanged between tube wall and flow: the flux is gained by the fluid $CV$

component $u$ depends solely on $r$. $\mathbf{w}_\infty$ relates to the nominal mass flow rate through the cross-section $\Omega$ so that

$$\dot{m} \equiv \rho_f \mathbf{w}_\infty \Omega \qquad (3.16 \text{ revisited})$$

The tube is very long, so that the wetted cylindrical surface will have a size $A \equiv 2\pi R \times \Delta x$. The fluid's velocity is subject to the transition from the value at centerline $u_{max}$, to the no-slip condition at the tube wall.

The temperature the fluid had before entering the tube is $T_\infty$. This tube surface is set and maintained at a *uniform temperature* $T_p$, with $T_p > T_\infty$. Correspondingly, the fluid is heated up from a centerline value $T_0$ to $T_p$: the heat gained by the fluid is represented by the uniform convective *thermal flux* $\dot{q}_h$, transferred from the tube wall and convected along with the fluid. In other words, the boundary conditions are

$$u(x, 0) = u_{max}, \quad u(x, R) = 0, \quad T(x, 0) = T_0, \quad T(x, R) = T_p \qquad (4.2)$$

Given the inherent geometry, rather than the centerline temperature $T_0$, the bulk space-averaged $\langle T \rangle$ is preferred to characterize the heat transfer. Coherently with the flow system definition in Thermodynamics, let $T_b$ be the cross-section weighted or bulk average of the local fluid temperature $T$, with the local velocity (flow-wise) component $u$ as the weighting factor:

$$T_b \equiv \langle T \rangle_b \equiv \frac{1}{\mathbf{w}_\infty \Omega} \int_\Omega u T \, d\Omega \qquad (4.3)$$

### 4.1.2.3 Link Between Flow Field and Convection Heat Transfer

We can complete now the assigned task that was initiated in Chap. 2, once a $CS$ and a $CV$ are assigned: the essential problems in studying convection heat transfer are

- the *determination of the thermal power transmitted under a given temperature difference and flow field* through the $CS$, and
- the *determination of the temperature distribution* on the $CS$ or in the $CV$ [1]

We already used the concept of **convective thermal flux** $\dot{q}_h$, that is, the thermal power $\dot{Q}_h$ transferred by convection through the exposed surface $A$, depending on the temperature difference or gradient $\Delta T \equiv T_p - T_\infty$ or $\Delta T \equiv T_p - T_b$ (for the situations represented in Figs. 4.2 and 4.3, respectively) and the inherent flow features. It is therefore:

$$\frac{\dot{Q}_h}{A} \equiv \dot{q}_h \propto \Delta T \tag{4.4}$$

The thermal flux $\dot{Q}_h/A$ causes a constant **net transport of heat between surface and fluid**, *in the direction taken by temperature decrease*. Differently than with the discussion of Eq. (2.2), now with the implicit action of fluid flow Eq. (2.2) evidences that the **temperature gradient is again the driving force of heat**: but as the temperature field is now intertwined with the distribution of velocity, Eq. (4.4) also illustrates the existing link between the *heat transferred to or from the fluid* and the *observable flow field*.

The distinctive proportionality parameter inherent to Eq. (4.4) is characteristic of the surface/fluid coupling, and it is called the **average** (in the sense of uniform along a given surface) **convective heat transfer coefficient** or **transmittance** $\overline{h}_T$:[2]

$$\boxed{\frac{\dot{Q}_h}{A} \equiv \overline{h}_T \Delta T} \tag{4.5}$$

with $\left[\overline{h}_T\right]$=W/m$^2$K (as we know already). This is the **Newton's Law of convection cooling**:[3] *the thermal power transferred by convection to or from a fluid flow is equal to the product of a heat transfer coefficient, the area of the surface through which the*

---

[1] In this case, the distribution of temperature depends on the previously determined distribution of velocity or may be even a source of it. See Sect. 4.1.2.4 (p. 150). Of course, the first problem is solved once the distribution of the temperature is ascertained, as we will see in the course of the chapter.

[2] The coefficient $\overline{h}$ was already introduced in Sect. 2.2.1 (p. 21) and employed in the development of conduction calculations, starting from Eq. (2.32). Here, to prepare the forthcoming discussion on the analogy between the *heat* transfer coefficient now at stake and a *mass* transfer coefficient, the subscript "T" is purposely added to refer to the former, while the subscript "M" will be used to refer to the later.

[3] As proposed, in a less explicit form, by the aforementioned scientist I. NEWTON, in the early eighteenth century.

*heat flows (measured perpendicularly to the direction of the flux), and a temperature gradient dependent on the given configuration.*

One can evaluate $\overline{h}_T$ by looking at the fluid side of either confining surfaces in Figs. 4.2 and 4.3: due to the no-slip condition at the walls, the *heat flux that traverses the laminar sublayer is ruled by sole conduction in the fluid* as seen in Sect. 3.5.2 (p. 115), whose thermal conductivity is $\lambda_f$. Then, we can apply to the fluid the balance

$$-\lambda_f \frac{dT}{dx} = \overline{h}_T \left( T - T_\infty \right) \qquad \text{(2.33b revisited)}$$

and use the geometry nomenclature of Figs. 4.2 and 4.3, so to have

$$\overline{h}_T = -\frac{\lambda_f}{T_p - T_\infty} \left. \frac{dT}{dy} \right|_{y=0^+} \qquad \text{for the external flow} \qquad \text{(4.6a)}$$

$$\overline{h}_T = -\frac{\lambda_f}{T_p - T_b} \left. \frac{dT}{dr} \right|_{r=R^-} \qquad \text{for the internal flow} \qquad \text{(4.6b)}$$

where superscript signs are supplied to provide the direction of the derivative calculation along the given coordinate axis. Note that $\overline{h}_T$ is always positive, as the sign of the thermal flux, positive when entering the $CV$ at stake, is assigned by the temperature difference, see Eq. (4.5).

With these definitions explicitly linked to the temperature derivative, the need of the temperature field solution in the fluid in order to evaluate the convective thermal flux is ascertained.

### 4.1.2.4   Operating Regimes and Bulk Flow Assumption

It is clear that fluid properties and velocity and temperature distributions *in the vicinity of the interacting surface* all have a strong influence on the determination of the heat exchange by convection. We will be soon examining in some detail the interaction between fluid dynamics and convection heat transfer, *in the boundary layer region* where its effects are found. Nevertheless, the definitions of Eqs. (4.6) are important as they help evaluate the **bulk heat transfer** and its OoM. In Table 4.1, few indicative values of $\overline{h}_T$, also depending on the inherent operating regime, are summarized.

Some of these regimes will be the subject of further scrutiny later on. It will suffice now to introduce the following operating regime definitions:

- *forced convection*: when fluid is forced (by a fan or a pump) to flow past the thermally interacting surface; these configurations are mostly present in many industrial processes
- *natural or free convection*: when the fluid flows past that surface as driven by the buoyancy or gravitation effect due to temperature differences in the fluid itself; these configurations are common in environmental situations

**Table 4.1** Indicative value ranges of the average convective heat transfer coefficient $\bar{h}_T$ (W/m$^2$K)

| Fluid, condition | W/m$^2$K |
|---|---|
| Air, in natural convection | 6–30 |
| Superheated steam or air, in forced convection | 30–300 |
| Oil, in forced convection | 60–1700 |
| Liquid water, in forced convection | $300–12 \times 10^3$ |
| Boiling water | $3 \times 10^3–60 \times 10^3$ |
| Saturated steam, in condensation | $6 \times 10^3–120 \times 10^3$ |

- *phase-change and multiphase convection*: when the fluid experiences condensation or boiling, or when multiple aggregation state are simultaneously present

The evaluation of the convective heat transfer coefficient is frequently the central task in convection analysis, via Eq. (4.5). The configurations at hand are seldom the simplified ones depicted in Figs. 4.2 and 4.3: the coefficients recalled here are *uniform* along and over the confining surfaces. It is clear that this fact is rather strong an idealization (due to the simplistic assumption of developed flow): indeed, **only bulk/uniform convective processes can be described by uniform convective coefficients**, due to flow deformations and thermal condition variations inherent in the flow/confining surface interaction, that consequently affect local heat transfer. So, *local effects in convective heat exchange cannot be described by such coefficients.*

## 4.2 Elementary Forced Convection: Heat Exchangers

A heat exchanger (HEX) is the device that enables forced convection between a fluid stream and a solid surface at different temperatures. Most of the times, heat exchange is accomplished by two streams[4] *in thermal contact* through a thin metal wall, while direct contact by mixing streams is also possible. HEXs are countless in the household, power productions, air treatments, chemical plants, and automotive sectors. HEX design involves a number of structural, regulation, and management aspects, as well as ancillary flow considerations such as load loss, and circulation pumps and fans.

HEX thermal design and verification can be studied with a macroscopical approach as fluid and heat respective flow along their one bulk direction. The total exchange surface of a HEX can be calculated when specific heats, flow rates, and inlet/outlet temperatures of the fluids are assigned, whereas for an existing HEX the transferred thermal power can be determined once the geometry, and fluid flow rates, specific heats, and inlet temperatures are referred. Depending on HEX type, we can

---

[4]Also called the *primary* or *working fluid*, and the *secondary* or *service fluid*.

recur to energy balances to describe the temperature progress and the thermal power transferred by such devices.

### 4.2.1   Classification of Heat Exchangers

HEXs are classified according to their scope and thermal contact, in relation to flow direction. The simplest configuration is the *single-stream* HEX, in which only one fluid is treated, as in *evaporators* or *condensers*.[5] With two different, changing temperature streams, many situations are possible. In the *parallel-flow* (cocurrent) configuration, the streams enter at the same end and travel in parallel with one another to the other end, while in the *counterflow* (countercurrent) configuration the streams enter from opposite ends (Fig. 4.4). The later arrangement is the more efficient than the former in transferring heat, due to the higher average temperature difference along any unit length along the device. In *cross-flow* HEXs, such as in Fig. 4.5, the streams are arranged roughly perpendicular to one another. The thermal contact can be replicated by stacking a base geometry in layers, to allow more flow rate and/or extend thermal contact.

Another HEX classification is based on construction type. Most commonly devised in *tubular* geometries, these devices come in many arrangements and sizes, the simplest one being the *coaxial* tube (also called a *shell*) sketched in Fig. 4.4: an example of *one-shell-pass, one-tube-pass*. *Internal baffles* can be inserted as in Fig. 4.6, to enhance mixing in the shell-pass side and provide a cross-flow component over the tube-passes.

Yet, an important HEX class is the *compact heat exchangers*. Thanks to multiple stacked elements, the heat transfer surface to gross size ratio is larger than with

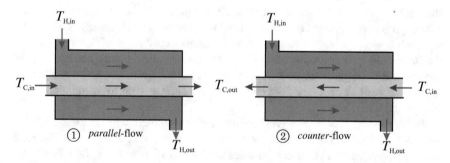

**Fig. 4.4** Simple HEX configurations, with colors and subscripts corresponding to relatively differ-ent (H/C) temperatures and direction (in/out) streams. ① *Parallel*-flow. ② *Counter*-flow

---

[5]In this cases, the nomenclature refers to the secondary fluid being at constant temperature, when undergoes phase-change or corresponds to a heat sink. Thus, the secondary fluid direction is imma-terial, and the primary fluid direction is irrelevant.

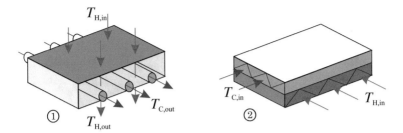

**Fig. 4.5** Two *cross*-flow heat exchanger, with colors and subscripts corresponding to relatively different (H/C) temperatures and direction (in/out) streams. ① The warmer stream is free to mix laterally in its flow through the device. ② The streams cannot mix laterally

**Fig. 4.6** A *two-shell-pass, four-tube-pass* baffled heat exchanger, with colors and subscripts corresponding to relatively different (H/C) temperatures and direction (in/out) streams

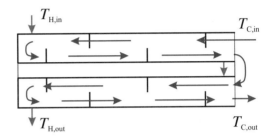

the other types.[6] They are commonly used when one of the two fluid is gas: on its side, many fins are adopted so to compensate for the low heat transfer coefficient associated with lower density fluid.

## 4.2.2 Fluid Temperature Progress

Associated with HEX type and stream arrangement is the characteristic temperature progress along the position $x$ along the effective length $L$, as shown in Fig. 4.7 in the assumptions of steady state and perfect insulation with the environment.

① In a condenser HEX, the inlet colder stream comes into thermal contact with the condensing fluid, at uniform warmer temperature $T_H$. As explained later, an evaporator HEX would exhibit a similar behavior.

② In a parallel-flow HEX, the temperature difference $(T_H - T_C)$ always decreases with increasing $x$. Notice that the colder $T_{C,out}$ at outlet cannot exceed the warmer $T_{H,out}$ at outlet. Also, notice that the last HEX segment (past a certain length $x'$ from inlet) is quite ineffective, and any HEX extension (longer $L$) would be worthless.

---

[6]Typically, several hundreds of square meters per cube meter.

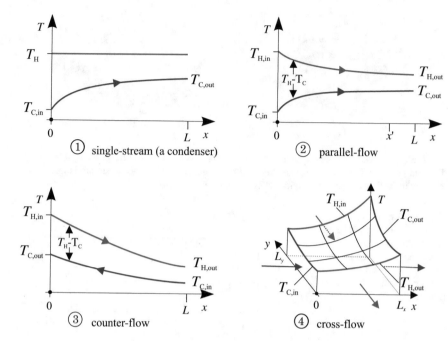

**Fig. 4.7** Characteristic fluid temperature progress for simple HEX configurations, from ① to ④, with colors and subscripts corresponding to relatively different (H/C) temperatures and direction (in/out) streams. For parallel- and counterflow cases, ② and ③, respectively, the local temperature difference $T_H - T_C$ is also denoted

③ In a counterflow HEX, the temperature difference ($T_H - T_C$) may increase or decrease (as in Figure) with increasing $x$ or, as a special case, it may be uniform. Notice that $T_{C,out}$ can well exceed $T_{H,out}$: for this reason, counterflow HEXs are preferred in the applications, except in single-stream configurations where the choice is irrelevant.

④ Finally, in a cross-flow HEX of effective width $L_x \times L_y$, the temperature patterns are more complex since the streams flow along two coordinates ($x$ and $y$), according to the internal flow arrangement.

Let us introduce the *flow thermal capacity C*, with $C \equiv \dot{m}c_p$; now, notice that the temperature progress is strongly dependent on the *ratio of the flow thermal capacities*:

$$\frac{C_H}{C_C} \equiv \frac{(\dot{m}c_p)_H}{(\dot{m}c_p)_C} \tag{4.7}$$

For example, compare the temperature progress for parallel- and counterflow HEXs in Fig. 4.7 ② and ③, that have $C_H/C_C$ respectively equal and less than 1, with the associated configurations in Fig. 4.8 ① and ②, featuring different values of $C_H/C_C$.

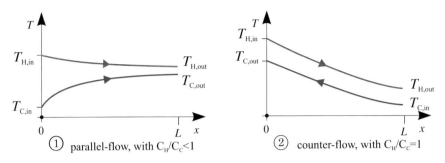

**Fig. 4.8** Characteristic fluid temperature progress for ① parallel-flow and ② counterflow HEXs, with colors and subscripts corresponding to relatively different (H/C) temperatures and direction (in/out) streams

### 4.2.3 Macroscopic Energy Balances and the Overall Heat Transfer Coefficient

In the above-mentioned assumptions, a *macroscopic energy balance* can be applied to the $CV$ enclosing any HEX shown so far. As for compact flow devices, the changes in kinetic and potential energy are negligible, and no work is exchanged. Therefore, we have in terms of *enthalpy flow rates*

$$(\dot{m}_H h_H + \dot{m}_C h_C)_{x=0} = (\dot{m}_H h_H + \dot{m}_C h_C)_{x=L}$$

Rearranging,

$$\dot{Q} = \dot{m}_H \left( h_{H,0} - h_{H,L} \right) = \dot{m}_C \left( h_{C,L} - h_{C,0} \right) \tag{4.8}$$

where $\dot{Q}$ is the thermal power transferred from the warmer to the cooler fluid. In case of a two-phase stream, its enthalpy difference is substituted by the related latent heat $\Delta h_V$, while in case of heat exchange toward a heat sink, only the balance on the fluid in thermal evolution will suffice.

If the specific heats $c_p$ can be assumed constant, then Eq. (4.8) is rewritten as

$$\dot{Q} = \left( \dot{m} c_p \right)_H \left( T_{H,0} - T_{H,L} \right) = \left( \dot{m} c_p \right)_C \left( T_{C,L} - T_{C,0} \right) \tag{4.9}$$

As with any balance equation, Eq. (4.9) can be used to find the value of an unknown parameter in the design dataset: for example, an outlet temperature can be calculated if the flow rates, the inlet temperatures, and the other outlet temperature are known, or the maximum possible thermal power transferred in a given situation. However, care must be exercised in order to obey to the Second Law of Thermodynamics, for heat always flows in the direction of temperature decrease.[7]

---

[7] As an example, we would want to heat some liquid water from $T_{C,0} = 30.0\,°C$ up to $T_{C,L} = 50.0\,°C$ with some hot combustion exhaust, the available stream being at $T_{H,0} = 180.0$ °C. The working

Most of the HEX types described so far involve convective heat transfer from one fluid to another across a plate or tube wall, the latter predominating. For these cases, the overall heat transfer coefficient $U$ was already introduced in Sects. 2.2.3 (p. 23) and 2.3.3.4 (p. 40), respectively. A $CS$ encasing one of the two fluids can be identified to be the heat exchange area $A$; with $P$ the internal wetted perimeter, it is $A \equiv PL$. With this definition, the sum of the resistances for the thermal circuit of Fig. 2.19 can be written again

$$\frac{1}{UA} = \frac{1}{\bar{h}_0 2\pi r_1 L} + \frac{\ln \frac{r_2}{r_1}}{2\pi \lambda_A L} + \frac{\ln \frac{r_3}{r_2}}{2\pi \lambda_B L} + \frac{1}{\bar{h}_\infty 2\pi r_3 L} \qquad (2.66)$$

and when $L$ is canceled throughout:

$$\frac{1}{UP} = \frac{1}{\bar{h}_0 2\pi r_1} + \frac{\ln \frac{r_2}{r_1}}{2\pi \lambda_A} + \frac{\ln \frac{r_3}{r_2}}{2\pi \lambda_B} + \frac{1}{\bar{h}_\infty 2\pi r_3} \qquad (4.10)$$

For the HEX element $\Delta x$ long, the exchanged thermal power is therefore

$$\Delta \dot{Q} = UP\Delta x \, (T_H - T_C) \qquad (4.11)$$

Using the product $UP$ allows one to disregard the actual value of $P$ and the corresponding values of $U$ and $(T_H - T_C)$ to employ (depending on the circuit segments as in Fig. 2.19), since Eq. (4.10) specifies the $UP$ product and not $U$ itself.

After a HEX has been in service for some time, *fouling* may form on any of the heat transfer surfaces, depending on the nature of the fluid. Examples include soot in furnaces, mineral deposits in water heaters, biochemical/biological strata or accumulation in HEXs used in biotechnology, and rust. According to the specific material and thickness, an additional resistance may be added in the thermal circuit and related Eq. (4.10), on the side where fouling occurs.

---

fluid has $\dot{m}_C = 5.0$ kg/s and $c_C = 4.20$ kJ/kgK, while the service fluid $\dot{m}_H = 10.0$ kg/s and $c_{pH} = 1.10$ kJ/kgK. On the water side, Eq. (4.9) gives

$$\dot{Q} = (\dot{m}c)_C \, (T_{C,L} - T_{C,0}) = 10.0 \times 4.20 \times 20.0 = 840 \text{ kW}$$

Reapplying Eq. (4.9) on the exhaust side:

$$T_{H,L} = T_{H,0} - \frac{\dot{Q}}{(\dot{m}c_p)_H} = 180.0 - \frac{840}{5.0 \times 1.10} = 27 \,^\circ\text{C}$$

which represents a violation of the Second Law of Thermodynamics, the temperature of the working fluid being always higher (regardless the HEX type and stream arrangement).

### 4.2.4 Heat Exchanger Analysis

HEX design is usually perform to maximize the surface area of the wall between the two streams, while minimizing flow resistance. HEX performance can also be affected by the addition of fins or corrugations in any of the streams, which increase surface area and may channel fluid flow or induce turbulence.

In Sect. 4.2.2 (p. 153), the progress of temperature along HEXs was anticipated. Now, we proceed to obtain their performance, based on simplified energy balances.

#### 4.2.4.1 Single-Stream HEXs

Figure 4.7 ① showed the progress of temperature along a condenser. Now we proceed by extending the analysis to an evaporator.

Corresponding to any of the configurations in Fig. 4.4, an evaporator HEX with its temperature progress is depicted in Fig. 4.9, a hot gas stream coming into thermal contact with saturated vapor at uniform cooler temperature $T_C$. If the latter is the working fluid, the HEX is also called a *boiler*.

A macroscopic energy balance sees that the needed latent heat of vaporization $\Delta h_v$ is transferred by convection from the hot gas to the inner tube wall, by conduction across the tube wall, and by a boiling process into the saturated vapor, whose temperature is $T_{sat}$ which depends on saturated vapor pressure $p$. As with Eqs. (4.8, 4.9) and related discussion, this balance yields

$$\left(\dot{m}c_p\right)_{\mathrm{H}}\left(T_{\mathrm{H,in}} - T_{\mathrm{H,out}}\right) = \left(\dot{m}\,\Delta h_v\right)_{\mathrm{C}} \tag{4.12}$$

With the knowledge of flow rate $\dot{m}_{\mathrm{H}}$ and inlet temperature $T_{\mathrm{H,in}}$ of the hot stream, Eq. (4.12) relates the outlet gas temperature, $T_{\mathrm{H,out}}$, to the amount of working gas fraction produced, $\dot{m}_{\mathrm{C}}$.

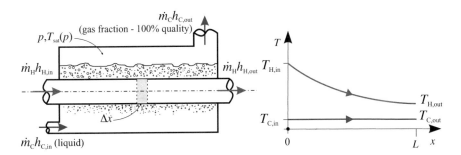

**Fig. 4.9** A single-tube evaporator. Left: the enthalpy flow rates crossing the HEX's $CS$, an elemental $CV$ of thickness $\Delta x$ of the inner tube, and indication of pressure $p$ in the saturated vapor; right: characteristic fluid temperature progress. Colors and subscripts correspond to different (H/C) temperatures and direction (in/out) streams

**Fig. 4.10** A single-tube
evaporator: close-up of the
elemental, axisymmetric
inner tube $CV$ in Fig.4.9,
left, with indication of the
thermal balance through its
$CS$. Notice that this $CV$
includes a portion of tube
wall

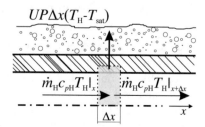

To obtain the variation of the service fluid temperature along the HEX $T_H(x)$
and its performance, we make an energy balance on an elemental $CV$ $\Delta x$ long, as
illustrated in Fig.4.10. Let us assume that the parasitic conduction along the inner
tube wall is small and can be neglected. The thermal power transferred across the
inner tube wall must equal the gas flow rate times its enthalpy decrease: for the gas
specific heat assumed to be constant

$$U P \Delta x \left[ T_H(x) - T_{sat} \right] = \left( \dot{m} c_p \right)_H \left( T_H|_x - T_H|_{x+\Delta x} \right) \qquad (4.13)$$

where $U$ and $P$ must be based on the same inner tube diameter, either the internal
or the external one. Dividing Eq. (4.13) by $\Delta x$, and taking the limit of its terms as
$\Delta x$ approaches 0, we rearrange

$$\frac{dT_H(x)}{dx} + \frac{U P}{\left( \dot{m} c_p \right)_H} (T_H - T_{sat}) = 0 \qquad (4.14)$$

The BC of this problem is of the first-kind:

$$T_H(x) = T_{H,in}, \text{ for } x = 0 \qquad (4.15)$$

Integrating and using Eq. (4.14) to evaluate the constant of integration gives

$$T_H(x) - T_{sat} = \left( T_{H,in} - T_{sat} \right) \exp - \left[ \frac{U P}{\left( \dot{m} c_p \right)_H} \right] x \qquad (4.16)$$

which exhibits an exponential decrease for $T_H(x)$, as anticipated by Fig.4.9, right.
The outlet gas temperature is obtained by letting $x = L$ in Eq. (4.16):

$$T_{H,out} - T_{sat} = \left( T_{H,in} - T_{sat} \right) \exp \left[ -\frac{U P L}{\left( \dot{m} c_p \right)_H} \right] \qquad (4.17)$$

Equation (4.17) can be arranged as

$$\frac{T_{H,in} - T_{H,out}}{T_{H,in} - T_{sat}} = 1 - \exp\left[-\frac{UPL}{(\dot{m}c_p)_H}\right] \tag{4.18}$$

or

$$\varepsilon = 1 - \exp(NTU) \tag{4.19}$$

Here, $\varepsilon$ is the HEX **effectiveness**, which represents the actual exchanged thermal power $\dot{Q} = (\dot{m}c_p)_H (T_{H,in} - T_{H,out})$ divided by the maximum possible thermal power that could be obtained in an infinitely long HEX. The dimensionless number NTU, the **Number of Transfer Units**, can be viewed as a measure of the HEX performance, as the larger is NTU, the higher the effectiveness $\varepsilon$. As design trade-off are to be considered, in practice values of $\varepsilon$ between 0.6 and 0.9 are typical. The same form of Eq. (4.19) would result in case on condenser analysis.

### 4.2.4.2 Double-Stream HEXs

In Fig. 4.7, the progress of temperature along parallel- and counterflow HEXs was anticipated, for cases ② and ③, respectively, along with their respective local temperature differences $T_H - T_C$. Now, we proceed by analyzing these double-stream HEXs. Recalling Eq. (4.11), the total thermal power transferred can be written as

$$\dot{Q} = UPL\overline{\Delta T} \tag{4.20}$$

with $\overline{\Delta T}$ a suitable mean temperature difference between the hot and cold streams.

To determine $\overline{\Delta T}$, we first consider a parallel-flow HEX, as illustrated in Fig. 4.11. Notice that these $CV$ include portions of both tube walls. Let us assume again that the parasitic conduction along the inner tube wall is small and can be neglected and that no external work is done and changes of kinetic energy and potential energy are negligible, as well. Application of the energy balance on the elemental $CV$ $\Delta x$ long containing either stream requires that

$$\Delta \dot{Q} = \dot{m}\Delta h = \dot{m}c_p\Delta T \tag{4.21}$$

**Fig. 4.11** A parallel-flow HEX. Elemental axisymmetric $CV$s with indication of the thermal power exchanged

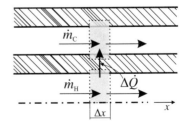

with $\Delta T \equiv T|_{x+\Delta x} - T|_x$. Looking at each $CV$ and applied thermal fluxes in Fig. 4.11, as the cold stream temperature is increasing along $x$, while the hot stream temperature is decreasing along $x$, it is, respectively,

$$U P \Delta x \left( T_H - T_C \right) = \left( \dot{m} c_p \right)_C \Delta T_C \tag{4.22a}$$

$$-U P \Delta x \left( T_H - T_C \right) = \left( \dot{m} c_p \right)_H \Delta T_H \tag{4.22b}$$

With the flow thermal capacity $C$ defined in Sect. 4.2.2 (p. 153), dividing Eqs. (4.22) by $\Delta x$ and taking the limit of their terms as $\Delta x$ approaches 0, we rearrange

$$C_C \frac{dT_C}{dx} = U P \left( T_H - T_C \right) \tag{4.23a}$$

$$C_H \frac{dT_H}{dx} = -U P \left( T_H - T_C \right) \tag{4.23b}$$

Subtracting Eq. (4.23a) from Eq. (4.23b) and further rearranging,

$$\frac{d \left( T_H - T_C \right)}{T_H - T_C} = -U P \left( \frac{1}{C_H} + \frac{1}{C_C} \right) dx$$

which can be integrated form $x = 0$ to $x = L$,

$$\ln \frac{T_{H,L} - T_{C,L}}{T_{H,0} - T_{C,0}} = - \int_0^L U P \left( \frac{1}{C_H} + \frac{1}{C_C} \right) dx \tag{4.24}$$

In this simplified presentation, we assume that $U P$ and specific heats are uniform along the HEX, and then

$$\ln \frac{T_{H,L} - T_{C,L}}{T_{H,0} - T_{C,0}} = -U P L \left( \frac{1}{C_H} + \frac{1}{C_C} \right) \tag{4.25}$$

Now, recalling Eq. (4.9)

$$\dot{Q} = \left( \dot{m} c_p \right)_H \left( T_{H,0} - T_{H,L} \right) = \left( \dot{m} c_p \right)_C \left( T_{C,L} - T_{C,0} \right) \tag{4.9}$$

it is also

$$C_H = \frac{\dot{Q}}{T_{H,0} - T_{H,L}}, \qquad C_C = \frac{\dot{Q}}{T_{C,L} - T_{C,0}} \tag{4.26}$$

Substituting back into Eq. (4.25) and rearranging gives

$$\dot{Q} = U P L \frac{\left( T_H - T_C \right)_L - \left( T_H - T_C \right)_0}{\ln \left[ \left( T_H - T_C \right)_L / \left( T_H - T_C \right)_0 \right]} \tag{4.27}$$

Comparing Eqs. (4.27) and (4.20) gives the required formula for $\overline{\Delta T}$:

$$\overline{\Delta T} \equiv \Delta T_{\text{lm}} \equiv \frac{(T_{\text{H}} - T_{\text{C}})_{\text{L}} - (T_{\text{H}} - T_{\text{C}})_0}{\ln\left[(T_{\text{H}} - T_{\text{C}})_{\text{L}} / (T_{\text{H}} - T_{\text{C}})_0\right]} \qquad (4.28)$$

$\Delta T_{\text{lm}}$ is called the **log mean temperature difference**, usually abbreviated as **LMTD**. Repeating this analysis for a counterflow HEX, the same result is obtained.

For double-stream HEXs other than the ideal coaxial type, a correction factor $F$ can be applied to the LMTD:

$$\dot{Q} = UPLF\Delta T_{\text{lm}} \qquad (4.29)$$

Compilations of $F$ factors exist for a variety of configurations and construction types.

## 4.3 Fundamental Equation of Thermal Convection

The definition of the convective heat transfer coefficient cannot run out the need to solve the thermal convection problem: it is clear that, in order to calculate $\dot{Q}_h$ with

$$\frac{\dot{Q}_h}{A} \equiv \overline{h}_{\text{T}}\Delta T \qquad (4.5)$$

due to the definition of the convective heat transfer coefficient $\overline{h}_{\text{T}}$, the velocity and temperature fields in the vicinity of the interacting surface must be determined. The analytical and numerical development of Chap. 3 will be sufficient for solving the velocity fields; in analogy with what has been carried out so far, to determine the distribution of temperature scalar $T$, it is therefore necessary to resort to a *generalized formulation*. This formulation is based on the balances carried out for the thermal conduction and the fluid mechanics, as seen in Sects. 2.3 (p. 25) and 3.3 (p. 87), respectively. To this end, let us consider again the $CV$ in Fig. 3.12.

### 4.3.1 Equation of Energy Conservation for a Fluid in Motion

Due to the simultaneous presence of heat and flow, an energy balance is can be written exploiting what we learned so far. Then, assuming for the time being *constant fluid thermodynamic properties*, let us apply the energy balance on the $CV$ in terms of *entering flux* (thermal power) through the $CS$, in the presence of mechanical work $W$ applied to the $CV$ *by the external forces*,[8] and in terms of *specific energy i*:

---

[8] The algebraic sign of $W$ is consistent with the sign convention adopted in engineering Thermodynamics textbooks.

$$\boxed{\text{net } \dot{Q}_\lambda \text{ through } CS} + \boxed{\text{net } \dot{W} \text{ at } CV} =$$

$$\boxed{\text{source/sink of } i \text{ in } CV, \text{ in the unit } \theta} + \boxed{\text{variation of } i \text{ in } CV, \text{ in the unit } \theta}$$

$$(4.30)$$

It is clear now that, as we encounter two different phases (solid s and fluid f), we will elaborate specific subscripts for the rheological properties, when deemed necessary. We proceed by developing these 4 terms:

1. in analogy with the development leading to the fundamental Equation of heat transfer by conduction, Eq. (2.21), we have

$$\boxed{\text{net } \dot{Q}_\lambda \text{ through } CS} = \lambda_f \nabla^2 T \mathrm{d}x \mathrm{d}y \qquad (4.31)$$

2. as discussed already when dealing with the balance of the Equation of momentum conservation

$$\boxed{\sum F_x \text{ applied on } CV} = \boxed{\text{variation of } mu \text{ in } CV, \text{ in the unit } \theta} \qquad (3.58)$$

the external forces inducing the mechanical power $\dot{W}$ acting on the $CV$ are subdivided in:

- *contact* or *surface forces* (as illustrated by Figs. 4.12 and 4.13);
- *mass forces* (as illustrated by Fig. 4.14).

Let us sum up these terms with their appropriate sign, as applied to $CS$ and $CV$, respectively:

$$\boxed{\text{net } \dot{W} \text{ at } CV} = \left[ \frac{\partial \left( u\sigma_{xx} \right)}{\partial x} + \frac{\partial \left( u\tau_{yx} \right)}{\partial y} + \frac{\partial \left( v\sigma_{yy} \right)}{\partial y} + \frac{\partial \left( v\tau_{xy} \right)}{\partial x} \right] \mathrm{d}x \mathrm{d}y$$
$$\pm u S_x \mathrm{d}x \mathrm{d}y \pm v S_y \mathrm{d}x \mathrm{d}y$$

$$(4.32)$$

3. in analogy with the development of the energy conservation Eq. (2.21), the source/sink is the energy flux due to any PCB phenomenon of interest:

$$\boxed{\text{source/sink of } i \text{ in } CV, \text{ in the unit } \theta} = \pm \dot{e}''' \mathrm{d}x \mathrm{d}y \qquad (4.33)$$

We encountered already, in Sect. 2.3.1 (p. 26), term $\dot{e}'''$ that accounts for any internal, uniform, generative or dissipative effects.

**Fig. 4.12** The infinitesimal
$CV$, interested by the
$x$-component contact force
fluxes through its $CS$

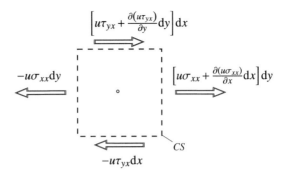

**Fig. 4.13** The infinitesimal
$CV$, interested by the
$y$-component contact force
fluxes through its $CS$

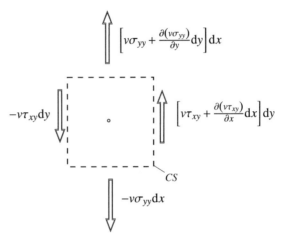

4. since we included the mass force term in expression for $\dot{W}$, the specific energy
   $i$ is composed by the remaining two terms: the specific internal energy $e$ and
   the specific kinetic energy $\frac{1}{2}\mathbf{w}^2 = \frac{1}{2}\mathbf{w}\cdot\mathbf{w} = (u^2 + v^2)/2$. Having assumed an
   *incompressible fluid*, we have therefore:

$$\boxed{\text{variation of } i \text{ in } CV, \text{ in the unit } \theta} = \rho_f \left[\frac{De}{D\theta} + \frac{D}{D\theta}\frac{(u^2 + v^2)}{2}\right] dx dy \quad (4.34)$$

Substitution of Eqs. (4.31–4.34) into the balance of Eq. (4.30) yields:

$$\lambda_f \nabla^2 T + \left[\frac{\partial\left(u\sigma_{xx}\right)}{\partial x} + \frac{\partial\left(u\tau_{yx}\right)}{\partial y} + \frac{\partial\left(v\sigma_{yy}\right)}{\partial y} + \frac{\partial\left(v\tau_{xy}\right)}{\partial x}\right]$$
$$\pm u S_x \pm v S_y = \pm\dot{e}''' + \rho_f\left[\frac{De}{D\theta} + \frac{D}{D\theta}\frac{(u^2 + v^2)}{2}\right] \quad (4.35)$$

After some manipulation, we can use Stokes' viscosity law, Eqs. (3.64), so that

**Fig. 4.14** The infinitesimal
$CV$, interested by the mass
force fluxes

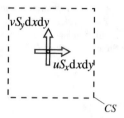

Eq. (4.35) becomes[9]

$$\rho_f \frac{De}{D\theta} = \lambda_f \nabla^2 T \pm \dot{e}''' + \mu\phi \tag{4.36}$$

with $\phi$ the *viscous dissipation function*,[10] defined by

$$\phi \equiv 2\left[\left(\frac{\partial u}{\partial x}\right)^2 + \left(\frac{\partial v}{\partial y}\right)^2\right] + \left(\frac{\partial u}{\partial y} + \frac{\partial v}{\partial x}\right)^2 \tag{4.37}$$

---

[9]Since it is generally $Dx^2 = 2xDx$, last term of Eq. (4.35) can be written as

$$\frac{\rho_f}{2}\frac{D}{D\theta}\left(u^2 + v^2\right) = \rho_f\left(u\frac{Du}{D\theta} + v\frac{Dv}{D\theta}\right)$$

The two material derivatives on the right-hand side will include the temporal and spatial variations
of the velocity components, so that Eqs. (3.63) can be employed to write:

$$\frac{\rho_f}{2}\frac{D}{D\theta}\left(u^2 + v^2\right) = u\frac{\partial\sigma_{xx}}{\partial x} + u\frac{\partial\tau_{yx}}{\partial y} \pm uS_x + v\frac{\partial\sigma_{yy}}{\partial y} + v\frac{\partial\tau_{xy}}{\partial x} \pm vS_y$$

With these terms devised, Eq. (4.35) is simplified:

$$\lambda_f \nabla^2 T + \left(\sigma_{xx}\frac{\partial u}{\partial x} + \tau_{yx}\frac{\partial u}{\partial y} + \sigma_{yy}\frac{\partial v}{\partial y} + \tau_{xy}\frac{\partial v}{\partial x}\right) = \pm\dot{e}''' + \rho_f\frac{De}{D\theta}$$

Now, we apply Stokes' viscosity law Eqs. (3.64) to resolve the normal and tangential stresses and
come up with pressure and velocity components:

$$\lambda_f \nabla^2 T + \left[-p\frac{\partial u}{\partial x} + 2\mu\left(\frac{\partial u}{\partial x}\right)^2 - p\frac{\partial v}{\partial y} + 2\mu\left(\frac{\partial v}{\partial y}\right)^2 + \mu\left(\frac{\partial u}{\partial y} + \frac{\partial v}{\partial x}\right)^2\right] = \pm\dot{e}''' + \rho_f\frac{De}{D\theta}$$

or

$$\lambda_f \nabla^2 T + \left[2\mu\left(\frac{\partial u}{\partial x}\right)^2 + 2\mu\left(\frac{\partial v}{\partial y}\right)^2 + \mu\left(\frac{\partial u}{\partial y} + \frac{\partial v}{\partial x}\right)^2\right] = \pm\dot{e}''' + \rho_f\frac{De}{D\theta}$$

having exploited the continuity Eq. (3.59).

[10]This function $\phi$ should be invoked when dealing with extremely viscous flows (thus with very
strong velocity gradients) such as in lubrication or oil piping problems. Expressions other than for
Cartesian 2-D space are available [1].

Recalling Note 10 of Chap. 2 on the use of specific heats, we have finally:

$$\rho_f c_{pf} \frac{DT}{D\theta} = \lambda_f \nabla^2 T \pm \dot{e}''' + \mu\phi \qquad (4.38)$$

that is the **Fundamental Equation of thermal convection**, in the assumption of *constant properties*. It can be shown [1] that Eq. (4.38) is a special form of the general temperature formulation with *variable properties*, also called the **First Law of Thermodynamics**, bearing an additional $\beta T \mathrm{D}p/\mathrm{D}\theta$ term:

$$\underbrace{\rho_f c_{pf} \frac{DT}{D\theta}}_{①} = \underbrace{(\nabla \cdot (\lambda_f \nabla T))}_{②} \underbrace{\pm \dot{e}'''}_{③} + \underbrace{\beta T \frac{Dp}{D\theta}}_{④} + \underbrace{\mu\phi}_{⑤} \qquad (4.39)$$

where $\beta$ is the *coefficient of thermal expansion*

$$\beta \equiv -\frac{1}{\rho_f} \left( \frac{\partial \rho_f}{\partial T} \right)_p \qquad (4.40)$$

The various terms appearing in Eq. (4.39) all represent thermal fluxes as well, *in competition* in every flow point: term ① represents the (transient) *convective flux*, term ② the *conduction flux*, term ③ the *internal heat source/sink*, term ④ the *portion of mechanical work W that is reversibly converted into heat by compression*, and finally term ⑤ the *portion of W that is irreversibly converted into heat by compression*.

Recalling the transfer of heat by conduction and the transfer of momentum, we find some **mathematical and physical analogies**, and some **differences** as well. In Eq. (4.39), the **transport and source terms** ① ② ③ are found again, while the **conversion-of-work-into-heat terms** ④ ⑤ are the *distinctive terms* for this transport mechanism.

Equation (4.39) is rarely used in this form. Besides Eq. (4.38) being valid for constant/uniform properties, other versions are usually encountered, for example, with *no viscous dissipation* and *source/sink terms* and for *constant conductivity*:

1. For an *ideal gas*, $\beta = 1/T$, and exploiting the relationship between specific heats $c_p - c_v = R$ (with $R$ the *constant of the specific gas*)[11] and the equation of state $p = \rho_f RT$ [2], it is therefore:

$$\rho_f c_v \left( \frac{\partial T}{\partial \theta} + (\mathbf{w} \cdot \nabla T) \right) = \lambda_f \nabla^2 T + p (\nabla \cdot \mathbf{w}) \qquad (4.41)$$

---

[11] As proposed by German physician and physicist J.R. VON MAYER, in mid-nineteenth century.

2. For a *fluid flowing at constant pressure*, it is $Dp/D\theta = 0$ and therefore:

$$\rho_f c_{pf} \left( \frac{\partial T}{\partial \theta} + (\mathbf{w} \cdot \nabla T) \right) = \lambda_f \nabla^2 T \qquad (4.42)$$

3. Most frequently, for a *fluid with constant density or incompressible*, $\beta = 0$ and the compressibility effect, term ④ in Eq. (4.38), is zero, and the viscous dissipation $\mu\phi$, term ⑤ in the same Eq. (4.38), is negligible: therefore, we have again:

$$\rho_f c_{pf} \left( \frac{\partial T}{\partial \theta} + (\mathbf{w} \cdot \nabla T) \right) = \lambda_f \nabla^2 T \qquad (4.43)$$

4. When trivially simplified, a last frequent form for *stationary media* and *uniform properties* can also be found:

$$\frac{1}{\alpha_s} \frac{\partial T}{\partial \theta} = \nabla^2 T \pm \frac{\dot{e}'''}{\lambda_s} \qquad (2.22 \text{ revisited})$$

*The system of the Navier–Stokes Equations*

$$\frac{\partial \rho_f}{\partial \theta} + (\nabla \cdot (\rho_f \mathbf{w})) = 0 \qquad (3.70a)$$

$$\rho_f \left[ \frac{\partial \mathbf{w}}{\partial \theta} + (\mathbf{w} \cdot \nabla) \, \mathbf{w} \right] = \mu \nabla^2 \mathbf{w} - \nabla p \pm \mathbf{S} \qquad (3.70b)$$

*and* (4.39) or one of its restricted forms, Eqs. (4.38, 4.41–4.43), together with the associated **initial conditions** and **boundary conditions**, *allows one to determine the functions* $T(x, y, \theta)$, $\mathbf{w}(x, y, \theta)$, and $p(x, y, \theta)$, in the specified assumptions.

It is evident that, with this fundamental Equation of thermal convection, all notations for BCs already specified in Sects. 2.3.1 (p. 26) and 3.3.4 (p. 94) may apply.

### 4.3.2   Solution Strategy

We have come a long way to get to this point, and the solution to convective heat transfer appears in its unveiled nature: the solution of $\mathbf{w}$ and $T$ can be found intertwined, depending on the case at stake. Indeed, prior to solution of governing Eq. (4.38) or the like, the governing Navier–Stokes Eqs. (3.70) above can be **independently** solved for $\mathbf{w}$ and $p$, **unless** the *fluid properties* $\mu$ and $\rho_f$, and *source/sink term* $\mathbf{S}$ depends on $T$.

In particular, as alluded to earlier in Sect. 2.1.3 (p. 17), a variation of the sole $\mu$ with $T$ can be dealt with by using numerical iterative techniques, as with the under-relaxation procedure introduced in Sect. 2.5.2.4 (p. 60). But a variation of $\rho_f$ due to temperature gradients implies the variation of **S** itself, as anticipated in Sect. 3.3.2 (p. 89): this is the important case of the *buoyancy force* (mass force induced by variation of fluid density) due to fluid temperature gradient. In this case, a **simultaneous solution** of Eqs. (3.70, 4.38) must be cast.

In any case it is well evident that this formidable system of governing Equations cannot be integrated with analytical techniques, except for few simplistic cases: that is why we turn again to computational methods, Sect. 4.6 (p. 192).

### 4.3.3 Other Coordinate Systems

In the same assumptions of Eq. (4.43), the energy conservation equation for the cylindrical geometry (Fig. 2.12) *in scalar notation* is reported here, for the sake of completeness:

$$
\boxed{
\begin{aligned}
\rho_f c_{pf} &\left( \frac{\partial T}{\partial \theta} + v_r \frac{\partial T}{\partial r} + \frac{v_\phi}{r} \frac{\partial T}{\partial \phi} + v_z \frac{\partial T}{\partial z} \right) = \\
&\lambda_f \left[ \frac{1}{r} \frac{\partial}{\partial r} \left( r \frac{\partial T}{\partial r} \right) + \frac{1}{r^2} \frac{\partial^2 T}{\partial \phi^2} + \frac{\partial^2 T}{\partial z^2} \right]
\end{aligned}
}
\tag{4.44}
$$

The energy conservation equation in a spherical coordinate system [2], that is, for the geometry illustrated in Fig. 2.13, is less frequently used.

### 4.3.4 Turbulent Heat Transfer

Like turbulent flow, turbulent heat transfer is encountered in many practical applications. In order to proceed along the line proposed in Sect. 3.3.6 (p. 98), in analogy with the development of the RANS in Sect. 3.3.7.3 (p. 101), we may consider that the measured temperature in one point in a given $CV$ of a turbulent flow will undergo some irregular, rapid fluctuation about a mean value. The actual temperature value $\overline{T}$ may be regarded as the sum of the mean (time-averaged) value and its fluctuation, and every consideration carried out in Sect. 3.3.7.3 will apply, to arrive to time-smoothed version of the Equation of thermal convection Eq. (4.43), in the assumption of *incompressible fluid*:

$$\left(\frac{\partial \overline{T}}{\partial \theta} + \frac{\partial \overline{T}\,\overline{u}}{\partial x} + \frac{\partial \overline{T}\,\overline{v}}{\partial y} + \frac{\partial \overline{T}\,\overline{w}}{\partial z}\right) + \underbrace{\left(\frac{\partial \overline{T'u'}}{\partial x} + \frac{\partial \overline{T'v'}}{\partial y} + \frac{\partial \overline{T'w'}}{\partial z}\right)}_{\text{turb. energy transport due to fluctuations}} = \alpha_{\mathrm{f}} \nabla^2 \overline{T}$$

$$(4.45)$$

(without the source/sink term, for the sake of simplicity). If we compare Eq. (4.43) with Eq. (4.45), beside the material derivative of the turbulent temperature $\overline{T}$, the under-braced terms appear, which represent **the energy transport associated with the turbulent fluctuations, which adds to the molecular transport**.

The eddy diffusivity concept is exploited again with the *thermal eddy diffusivity* $\varepsilon_\alpha$ (with $[\varepsilon_\alpha]=\mathrm{m}^2/\mathrm{s}$),[12] so to evidence the added contribution of turbulence to molecular heat transport. Defining the *apparent turbulent heat flux*

$$-\overline{T'w'_j} \equiv \varepsilon_\alpha \frac{\partial \overline{T}}{\partial x_j} \tag{4.46}$$

Thermal convection Equation (4.45) becomes

$$\boxed{\frac{\partial \overline{T}}{\partial \theta} + \frac{\partial}{\partial x_j}\left(\overline{T}\,\overline{w}_j\right) = \frac{\partial}{\partial x_j}\left((\alpha_{\mathrm{f}} + \varepsilon_\alpha)\frac{\partial \overline{T}}{\partial x_j}\right)} \tag{4.47}$$

Unlike molecular thermal diffusivity $\alpha_{\mathrm{f}}$, which is a parameter of the fluid, the thermal eddy diffusivity $\varepsilon_\alpha$ depends strongly on the position within the flow, while going to zero at the flow-limiting wall. Its value is approximately the same as the momentum eddy diffusivity $\varepsilon_\nu$ we dealt with in Sect. 3.3.7.4 (p. 104).

## 4.4   Dimensionless Equations and Phenomenology of Convection

The concepts presented in Sects. 2.4 (p. 49) and 3.4 (p. 108) on the dimensionless equations can be exercised again to identify the controlling dimensionless parameters that are commonly used in the applications of convection heat transfer. Moreover, an introduction is provided to analytical solutions and correlations for this mechanism. Since the notations differ depending on the convection regime, forced and natural convection will be reported on separately.

---

[12]It is worth to note the complete analogy with the definition of momentum eddy diffusivity $\varepsilon_\nu$ Eq. (3.92).

### 4.4.1 Forced Convection

Let us start by invoking the governing equations

$$(\nabla \cdot \mathbf{w}) = 0 \tag{3.70a revisited}$$

$$\rho_f \frac{D\mathbf{w}}{D\theta} = \mu \nabla^2 \mathbf{w} - \nabla p \tag{3.70b revisited}$$

$$\rho_f c_{pf} \frac{DT}{D\theta} = \lambda_f \nabla^2 T + \mu \phi \tag{4.38 revisited}$$

for an *incompressible fluid* with *constant rheological properties* and *in the steady state*, but *with no source/sink term*, to solve for $T = T(x, y)$, $\mathbf{w} = \mathbf{w}(x, y)$, and $p = p(x, y)$ (for the general $CV$ of Fig. 3.12). We were already able to examine the development of the fluid dynamic and the thermal boundary layers, as shown in Figs. 4.2 and 4.3. Then, let us choose some suitable dimensionless independent and dependent variables, relative to a flow having reference temperature, velocity, and pressure of $T_\infty$, $\mathbf{w}_\infty$, and $\rho_f \mathbf{w}_\infty^2$, respectively; the flow interacts with a generic geometry having a reference length $L$ and features an arbitrary reference temperature difference $(T_p - T_\infty) \equiv \Delta T$. We can choose therefore:

$$x^* \equiv \frac{x}{L}, \ y^* \equiv \frac{y}{L}, \ u^* \equiv \frac{u}{w_\infty}, \ v^* \equiv \frac{v}{w_\infty}, \ p^* \equiv \frac{p}{\rho_f w_\infty^2}, \ T^* \equiv \frac{T - T_\infty}{\Delta T} \tag{4.48}$$

We proceed by operating the change of variable as performed already in Sects. 2.4.2 and 3.4, that is, by substituting variables of Eqs. (4.48) in Eqs. (3.70, 4.38), to obtain again the dimensionless Navier–Stokes Equations, Eqs. (3.106), as well as the following **dimensionless energy conservation Equation**:

$$\begin{aligned}
u^* \frac{\partial T^*}{\partial x^*} + v^* \frac{\partial T^*}{\partial y} &= \frac{1}{Pe}\left(\frac{\partial^2 T^*}{\partial x^{*2}} + \frac{\partial^2 T^*}{\partial y^{*2}}\right) \\
&+ \frac{Ec}{Re_L}\left[2\left(\frac{\partial u^*}{\partial x^*}\right)^2 + 2\left(\frac{\partial v^*}{\partial y^*}\right)^2 + \left(\frac{\partial u^*}{\partial y^*} + \frac{\partial v^*}{\partial x^*}\right)^2\right]
\end{aligned} \tag{4.49}$$

Here, beside the Reynolds number

$$Re_L \equiv \frac{\rho_f w_\infty L}{\mu} \tag{3.2 revisited}$$

the following new *dimensionless numbers* have been defined:[13]

- **Prandtl number** that characterizes *the physical properties of a fluid with convective and diffusive heat transfer*:

$$\boxed{\mathrm{Pr} \equiv \frac{\mu c_{pf}}{\lambda_f} = \frac{\nu}{\alpha_f}}$$

(4.50)

involving the kinematic viscosity[14]

$$\nu \equiv \frac{\mu}{\rho_f}$$

(3.6)

Moreover, another Prandtl number can be invoked, in case of turbulent heat transfer: just as with Eq. (4.50), the ratio of the two eddy diffusivities $\varepsilon_\nu$ and $\varepsilon_\alpha$

$$-\overline{w_i' w_j'} \equiv \varepsilon_\nu \frac{\partial \overline{w}_i}{\partial x_j}$$

(3.91 revisited)

$$-\overline{T' w_j'} \equiv \varepsilon_\alpha \frac{\partial \overline{T}}{\partial x_j}$$

(4.46)

forms the *turbulent Prandtl number*

$$\boxed{\mathrm{Pr}_t \equiv \frac{\varepsilon_\nu}{\varepsilon_\alpha}}$$

(4.51)

From what was said in Sects. 3.3.7.4 and 4.3.4 (p. 104 and 167, respectively), its value is approximately equal to 1.

- **Peclet number** that is the criterion for the *mutual action of the convective and molecular heat transfers in flowing fluid*, since it expresses the ratio of the convective thermal flow transferred by the fluid (fluid enthalpy change $\Delta h$, flowing in the axial direction) to that transferred by conduction in the normal direction ($\dot{q}_{\lambda_f}$):[15]

$$\boxed{\mathrm{Pe}_L \equiv \mathrm{Re}_L \mathrm{Pr} \equiv \frac{w_\infty L}{\alpha_f}}$$

(4.52)

involving the definition of thermal diffusivity

---

[13]The Pr and Pe numbers were proposed in the early nineteenth century by the aforementioned scientists L. PRANDTL and J.C.E. PÉCLET, respectively, while the Ec number was were proposed by Czech-German origin American engineer E.R.G. ECKERT, in mid-twentieth century.

[14]Recall Note 10 in Chap. 3, on kinematic viscosity and thermal diffusivity definitions.

[15]The Pe number is of particular interest here: with the Pe number increasing, the heat conduction portion decreases and the convective heat portion grows.

$$\alpha_f \equiv \left. \frac{\lambda}{\rho c_p} \right|_f \qquad \text{(2.5 revisited)}$$

- **Eckert number** that expresses *the ratio of kinetic energy to a thermal energy change*:

$$\boxed{\mathrm{Ec} \equiv \frac{\mathbf{w}_\infty^2}{c_{pf}\,\Delta T}} \qquad (4.53)$$

In the end, we see that the functions **in dimensional form** we sought were

$$\mathbf{w} = \mathbf{w}\,(x, y, L, \rho_f, \mu, \mathbf{w}_\infty) \text{ and } p = p\,(x, y, L, \rho_f, \mu, \mathbf{w}_\infty) \qquad (3.105)$$

$$T = T\left(x_j, L, \alpha_f, T_\infty, \Delta T, \mathbf{w}_\infty\right) \qquad (4.54)$$

while we have obtained, beside

$$\mathbf{w}^* = \mathbf{w}^*\left(x^*, y^*, \mathrm{Re}_L\right) \text{ and } p^* = p^*\left(x^*, y^*, \mathrm{Re}_L\right) \qquad (3.106)$$

the following solution, also **in dimensionless form**:

$$\boxed{T^* = T^*\left(x^*, y^*, \mathrm{Re}_L, \mathrm{Pr}, \mathrm{Ec}\right)} \qquad (4.55)$$

with the resulting ease of discussion.

As for the appropriate BCs, a specific relationship must be supplemented, for the thermal contact or *interaction* between a fluid and a solid surface is now at stake. For example, we can manipulate the notation of average convective heat transfer coefficient for an external flow such as the one in Fig. 4.2 (but other configurations can be exercised, as well)

$$\overline{h}_T = -\frac{\lambda_f}{T_0 - T_\infty} \left. \frac{dT}{dy} \right|_{y=0^+} \qquad (4.6a)$$

by using the dimensionless definitions of Eqs. (4.48) to come up with

$$\boxed{\begin{aligned} \left|\overline{\mathrm{Nu}_L}\right| &\equiv \frac{\overline{h}_T L}{\lambda_f} \qquad &(4.56a) \\[2mm] \overline{\mathrm{Nu}_L} &= -\left. \frac{\partial T^*}{\partial y^*} \right|_{y^*=0^+} \qquad &(4.56b) \end{aligned}}$$

which define the **average**[16] **Nusselt number** (based on length $L$), which compares *the convective heat transfer to the conduction heat transfer.*[17]

Note that in Eq. (4.56a) the absolute value of this number has been taken, as the right-hand side is always inherently positive, while it is evident from Eq. (4.56b) that $\overline{\text{Nu}}_L$ can generally assume any sign, depending on the derivative at the right-hand side.

From Eq. (4.56b), it is evident that it will be

$$\overline{\text{Nu}}_L = \overline{\text{Nu}}_L \, (\text{Re}_L, \text{Pr}, \text{Ec}) \tag{4.57}$$

or, if the Ec/Re ratio is sufficiently small (as commonly found):

$$\boxed{\overline{\text{Nu}} = \overline{\text{Nu}} \, (\text{Re}_L, \text{Pr})} \tag{4.58}$$

We will be dealing soon with the evaluation of this paramount dimensionless number, occurring due to solid/fluid interaction.

### 4.4.2  Natural Convection

In this case, we proceed by assuming an *incompressible fluid* with *constant rheological properties*, but *including the source term* in the momentum equation this time. First of all, we soon realize that **an exception must be made with respect to density** $\rho_f$, for its thermal variation dictates the mass force which is the driving phenomenon now at stake.

So, let us consider a laminar flow in the steady state, as in Fig. 4.15, whose *upward motion* along a *vertical* plate is due to the *temperature difference* with the plate. As just said, it is precisely the temperature difference that drives, through the local decrease of fluid density along $y$, the upward motion and the development of fluid dynamic and thermal boundary layers. It is easy to imagine that a converse behavior will be obtained for a *cooled fluid in downward motion.*

The flow is again very wide along the other two coordinates. This time, the *undisturbed* velocity will be zero sufficiently far from the plate, outside the resulting boundary layer ($v_\infty = 0$): the fluid's velocity is subject to the transition from this zero value, to a maximum value just in the more dynamically active boundary layer, to the no-slip condition $v = 0$, again, at the plate.

The *undisturbed temperature* in the quiescent fluid away from the plate is $T_\infty$. This exposed surface is set and maintained at a *uniform temperature* $T_p$, with $T_p > T_\infty$. Correspondingly, the fluid is heated up from $T_\infty$ to $T_p$: the heat gained by the fluid

---

[16]This "average" adjective is purposely stressed-out here as this dimensionless Nu number depends on the definition of *average heat transfer coefficient* Eq. (4.6a).

[17]As proposed by German engineer E.K.W. NUSSELT in the early twentieth century.

**Fig. 4.15** Velocity and
temperature profiles in a
fully developed flow along a
heated vertical plate. The
infinitesimal $CV$ is provided,
along with the indication of
the gravity vector **g**, and the
mass force $S_y$ aligned with it

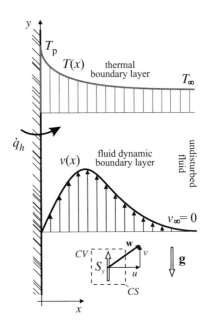

is represented by the uniform convective *thermal flux* $\dot{q}_h$, transferred from the plate
and convected along with the fluid. In other words, the BCs are

$$v(0, y) = 0, \quad v(\infty, y) = 0, \quad T(0, y) = T_{\mathrm{p}}, \quad T(\infty, y) = T_\infty \qquad (4.59)$$

In the steady state, let us start with the governing Equation

$$\rho_{\mathrm{f}} \left( u \frac{\partial v}{\partial x} + v \frac{\partial v}{\partial y} \right) = \mu \left( \frac{\partial^2 v}{\partial x^2} + \frac{\partial^2 v}{\partial y^2} \right) - \frac{\partial p}{\partial y} + S_y \qquad (3.69b)$$

where the appropriate source term $S_y = -\rho_{\mathrm{f}} g$ is included, with the sign referring to
the situation in Fig. 4.15. Moreover, we add the governing Equation for thermal convection to the analysis, with the exclusion of its source and work-into-heat conversion
terms

$$\rho_{\mathrm{f}} c_{pf} \left( u \frac{\partial T}{\partial x} + v \frac{\partial T}{\partial y} \right) = \lambda_{\mathrm{f}} \nabla^2 T \qquad (4.38 \text{ revisited})$$

If we assume that the results of the scale analysis of Sect. 3.5.1 (p. 110), with particular
reference to the discussion leading to

$$\frac{\partial^2 u / \partial x^2}{\partial^2 u / \partial y^2} \sim \left( \frac{\delta}{x} \right)^2 \ll 1 \qquad (3.111)$$

that is, negligible diffusive gradients along the plate for a "thin" boundary layer, still hold, so we end up with the governing Equation of energy

$$u\frac{\partial T}{\partial x} + v\frac{\partial T}{\partial y} = \alpha_f \frac{\partial^2 T}{\partial x^2} \tag{4.60}$$

and the governing Equation of momentum

$$\rho_f \left( u\frac{\partial v}{\partial x} + v\frac{\partial v}{\partial y} \right) = \mu\frac{\partial^2 v}{\partial x^2} - \frac{dp}{dy} - \rho_f g \tag{4.61}$$

which needs some additional term transformation. Far from the wall, velocity gradients are zero, so Eq. (4.61) reduces to

$$-\frac{dp}{dy} = \rho_{f\infty} g \tag{4.62}$$

Substituting back in Eq. (4.61) gives

$$\rho_f \left( u\frac{\partial v}{\partial x} + v\frac{\partial v}{\partial y} \right) = \mu\frac{\partial^2 v}{\partial x^2} + g(\rho_{f\infty} - \rho_f) \tag{4.63}$$

Now, the Equation of state $\rho_f = \rho_f(p, T)$ can be expanded following Taylor's series (see Note 22 of Chap. 2) about $T_\infty$ and $p_\infty$ to yield

$$\rho_f(T, p) = \rho_{f\infty} + \left(\frac{\partial \rho_f}{\partial T}\right)_p (T - T_\infty) + \left(\frac{\partial \rho_f}{\partial p}\right)_T (p - p_\infty) + \text{higher-order terms} \tag{4.64}$$

We can take, across the boundary layer, that pressure is almost uniform: in this way, we assume *constant rheological properties except for the density variation with temperature in the buoyancy term*: the so-called *Boussinesq approximation*.[18] Then, $p \approx p_\infty$ and if we further recall the definition of the coefficient of thermal expansion $\beta$

$$\beta \equiv -\frac{1}{\rho_f} \left(\frac{\partial \rho_f}{\partial T}\right)_p \tag{4.40}$$

we obtain

$$\frac{\rho_{f\infty} - \rho_f}{\rho_f} = -\frac{1}{\rho_f} \left(\frac{\partial \rho_f}{\partial T}\right)_p (T - T_\infty) = \beta(T - T_\infty) \tag{4.65}$$

Equation (4.63) is therefore simplified as

---

[18] As proposed by the aforementioned scientist V.J. BOUSSINESQ at beginning twentieth century.

$$u\frac{\partial v}{\partial x} + v\frac{\partial v}{\partial y} = \underbrace{\nu\frac{\partial^2 v}{\partial x^2}}_{②} \underbrace{+g\beta(T - T_\infty)}_{③} \tag{4.66}$$

$$\underbrace{\phantom{u\frac{\partial v}{\partial x} + v\frac{\partial v}{\partial y}}}_{①}$$

The three terms appearing in Eq. (4.66) hold the meaning of the *inertia or deceleration force* of the fluid $CV$, term ①, *viscous or friction force* transmitted to the fluid $CV$ by the solid surface, term ②, and *source/sink due to the thermal buoyancy force* transmitted to the fluid $CV$ by the inherent thermal distribution, term ③.

Now, our solving system made up of governing Equations (4.60, 4.66) supplemented by

$$\frac{\partial u}{\partial x} + \frac{\partial v}{\partial y} = 0 \tag{3.57}$$

must be solved for $T = T(x, y)$ and $\mathbf{w} = \mathbf{w}(x, y)$. Then, let us choose some suitable dimensionless independent and dependent variables, relative to a flow having reference temperature and velocity of $T_\infty$ and $\mathbf{w}_\infty$, respectively; the flow interacts with a generic geometry having a reference length $L$ and features an arbitrary reference temperature difference $(T_p - T_\infty) \equiv \Delta T$. We can choose therefore:

$$x^* \equiv \frac{x}{L}, \quad y^* \equiv \frac{y}{L}, \quad u^* \equiv \frac{u}{\mathbf{w}_\infty}, \quad v^* \equiv \frac{v}{\mathbf{w}_\infty}, \quad T^* \equiv \frac{T - T_\infty}{\Delta T} \tag{4.67}$$

We start by operating the change of variable as performed already in Sect. 3.4 (p. 108) for the Equation of continuity, that is, by substituting variables of Eqs. (4.67) in Eq. (3.57), to obtain again

$$\frac{\partial u^*}{\partial x^*} + \frac{\partial v^*}{\partial y^*} = 0 \tag{3.104a}$$

Then, we proceed with the momentum conservation Equation: notice that Re (based on a generic $\mathbf{w}_\infty$) in the present case is not an independent parameter influencing $T^*$, since no flow region where this imposed velocity exists, differently than with the BCs of the dimensional solution for forced convection

$$T = T\left(x, y, \rho_f, c_{pf}, \lambda_f, L, \mathbf{w}_\infty, T_\infty, \Delta T\right) \tag{4.54}$$

Therefore, we employ a reference velocity value based on the kinematic viscosity, instead:

$$\mathbf{w}_\infty \equiv \frac{\nu}{L} \tag{4.68}$$

Based on this, a Re number

$$\mathrm{Re}_L \equiv \frac{\mathbf{w}_\infty L}{\nu} \tag{3.2 revisited}$$

would be equal to 1. So, when we proceed further as in Sect. 3.4 with the remaining Equation of conservation, that is, by substituting variables of Eqs. (4.67) in Eqs. (4.66, 4.60), we obtain the following **dimensionless momentum and energy conservation Equations**:

$$u^* \frac{\partial v^*}{\partial x^*} + v^* \frac{\partial v^*}{\partial y} = \mathrm{Gr}_L T^* + \frac{\partial^2 v^*}{\partial x^{*2}} \tag{4.69a}$$

$$u^* \frac{\partial T^*}{\partial x^*} + v^* \frac{\partial T^*}{\partial y} = \frac{1}{\mathrm{Re}_L \mathrm{Pr}} \frac{\partial^2 T^*}{\partial x^{*2}} \tag{4.69b}$$

Here, beside the Prandtl number, already introduced by Eq. (4.50), the following *dimensionless number* has been defined:[19]

- **Grashof number**, or the *square of the Re number based on an equivalent velocity* $\mathbf{w}_{eq} = g\beta(T_0 - T_\infty)L$ *induced by buoyancy*:

$$\mathrm{Gr}_L \equiv \frac{g\beta \left(T_p - T_\infty\right) L}{\mathbf{w}_\infty^2} = \frac{g\beta \left(T_p - T_\infty\right) L^3}{v^2} = \left(\frac{\mathbf{w}_{eq} L}{v}\right)^2 \equiv \mathrm{Re}_{eq}^2 \tag{4.70}$$

In the end, we see that the function we sought was, **in dimensional form**:

$$\mathbf{w} = \mathbf{w}\,(x,\, y,\, L,\, v,\, \beta,\, T,\, T_\infty) \text{ and } T = T\,(x,\, y,\, L,\, \alpha_f,\, \mathbf{w}) \tag{4.71}$$

while we have obtained, beside Eqs. (3.108), **in dimensionless form**:

$$\mathbf{w}^* = \mathbf{w}^* \left(x^*,\, y^*,\, T^*,\, \mathrm{Gr}_L\right) \text{ and } T^* = T^* \left(x^*,\, y^*,\, \mathbf{w}^*,\, \mathrm{Pr}\right) \tag{4.72}$$

In fact, the role of Gr as an appropriate dimensionless group has been criticized,[20] but for the time being we have formally verified the intertwined nature of this analytical solution.

As for the appropriate BCs, similar to the notations leading to Eq. (4.71), a specific notation must be supplemented, for the thermal contact between a fluid and a solid surface. For example, we can manipulate again the definition of $\bar{h}_T$

$$\bar{h}_T = -\frac{\lambda_f}{T_0 - T_\infty} \frac{dT}{dy}\bigg|_{y=0^+} \tag{4.6a}$$

by using the dimensionless definitions of Eqs. (4.67) to come up again with

$$\left|\overline{\mathrm{Nu}_L}\right| \equiv \frac{\bar{h}_T L}{\lambda_f} \tag{4.56a}$$

---

[19] As proposed by German engineer F. GRASHOF, in mid-nineteenth century.

[20] See the discussion by Bejan, A.: Convection Heat Transfer. pp. 188–192. John Wiley & Sons, New York (1995). The topic is resumed later in Note 14 of Chap. 5.

$$\overline{\mathrm{Nu}}_L = - \left.\frac{\partial T^*}{\partial x^*}\right|_{x^*=0^+} \qquad \text{(4.56b revisited)}$$

which is the usual average Nusselt number comparing the convective heat transfer to the conduction heat transfer. From its definition, it is evident that it will be

$$\overline{\mathrm{Nu}}_L = \overline{\mathrm{Nu}}_L\,(\mathrm{Gr}_L, \mathrm{Pr}) \qquad (4.73)$$

or, with the following *dimensionless number*:[21]

- **Rayleigh number**, or the *product of Gr and Pr numbers*:

$$\boxed{\mathrm{Ra}_L \equiv \frac{g\beta\,(T_\mathrm{p} - T_\infty)\,L^3}{\nu\alpha_\mathrm{f}} = \mathrm{Gr}_L\mathrm{Pr}} \qquad (4.74)$$

At the end, we come up with the most simple functional dependence:

$$\boxed{\overline{\mathrm{Nu}}_L = \overline{\mathrm{Nu}}_L\,(\mathrm{Ra}_L)} \qquad (4.75)$$

As anticipated with the with forced convection, we will deal next with the evaluation of this paramount dimensionless number occurring due to solid/fluid interaction.

## 4.5   Convection Interactions: The Thermal Boundary Layer

We anticipated already in Sect. 3.5 (p. 109) on the importance of the boundary layer due to solid/fluid interactions in the laminar regime. It is clear that the deformation of heat carrier flows in a variety of processes yields temperature gradients and hence convective heat transfer to/from these surfaces. Reference to elementary configurations was already made in Sect. 3.2 (p. 74). With the ground set with the dimensionless analysis, we can determine some notations useful to convection heat transfer, but first we need to extend the physics described when studying the fluid dynamics, as the underlying concept to convective heat transfer is the temperature boundary layer.

### 4.5.1   Scale Analysis, Thicknesses Relationships

With Figs. 4.2, 4.3, and 4.15, we illustrated already both fluid dynamic and thermal boundary layers, for forced external/internal and natural convection situations, respectively, with uniform surface temperature and constant free-stream temperature. We may continue the analysis started in Sect. 3.5.1, to relate dimensionless

---

[21] As proposed by British physicist Lord Rayleigh, J.W. STRUTT, in the late nineteenth century.

groups with the onset of the fluid/thermal interaction between a laminar *thermally developing flow* and a perturbing surface.

Let us consider again the flow in Fig. 4.2 and revisited in Figs. 4.16 or 4.17 where the onset of a thermal boundary layer is now superimposed. The local convective thermal flux $\dot{q}_{hx}$ exchanged is reported only indicatively, as the surface is still supposed at *uniform temperature* and with the free stream at constant temperature, for the time being; it will be, however

$$\dot{q}_h x = \int_0^x \dot{q}_{hx} \, dx \tag{4.76}$$

At the steady state, as the deforming action due to fluid viscosity propagates normally to the plate yielding for a perturbed flow region with its thickness $\delta(x)$, in the same

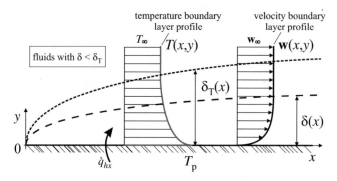

**Fig. 4.16** Flow parallel to a plate: the temperature and velocity distributions, $T(x, y)$ and $\mathbf{w}(x, y)$, being deformed along $x$, in the developing the boundary layer. The boundary layers are indicated by the $x$−varying thicknesses $\delta_T$ and $\delta$, with $\delta < \delta_T$. The thermal flux $\dot{q}_{hx}$ exchanged locally between plate and flow is also indicated

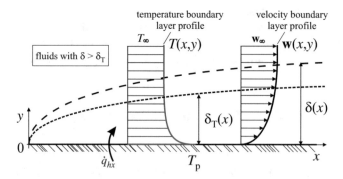

**Fig. 4.17** Flow parallel to a plate: the temperature and velocity distributions, $T(x, y)$ and $\mathbf{w}(x, y)$, being deformed along $x$, in the developing the boundary layer. The boundary layers are indicated by the $x$−varying thicknesses $\delta_T$ and $\delta$, with $\delta > \delta_T$. The thermal flux $\dot{q}_{hx}$ exchanged locally between plate and flow is also indicated

way the temperature $T(x, y)$ changes from its undisturbed value $T_\infty$ to $T_p$ of the surface, yielding for a thermally perturbed flow region with its thickness $\delta_T(x)$.

Once again, laminar flow regime will occur below certain velocity magnitude/travelled distance thresholds, as indicated by the local Reynolds number

$$\mathrm{Re}_x \equiv \frac{\mathbf{w}_\infty x}{\nu} \qquad\qquad (3.107 \text{ revisited})$$

In order to illustrate the dependence of thermal boundary layer features such as *thickness* and *driving temperature difference* on the heat exchange, it is appropriate to perform a scale analysis on governing Eq. (4.38) or (4.43), in the usual assumption of steady state and *incompressible flow with constant rheological properties*, and with no internal heat source/sink terms.

Let us assume that thickness $\delta_T$ (Figs. 4.16 and 4.17) is always very small when compared to $x$:

$$\boxed{\delta_T \ll x} \qquad\qquad (4.77)$$

In other words, we assume that **the thermal boundary layer is thin**. Based on a reasoning analogous to the one leading to development of

$$\frac{\partial^2 u/\partial x^2}{\partial^2 u/\partial y^2} \sim \left(\frac{\delta}{x}\right)^2 \ll 1 \qquad\qquad (3.111)$$

we conclude that in the governing Equation

$$\rho_f c_{pf} \left( \frac{\partial T}{\partial \theta} + (\mathbf{w} \cdot \nabla T) \right) = \lambda_f \nabla^2 T \qquad\qquad (4.43)$$

the $\partial^2 T/\partial x^2$ term is negligible and the steady-state problem reduces to

$$\underbrace{u \frac{\partial T}{\partial x} + v \frac{\partial T}{\partial y}}_{①} = \underbrace{\alpha_f \frac{\partial^2 T}{\partial y^2}}_{②} \qquad\qquad (4.78)$$

The two terms appearing in Eq. (4.78) hold the meaning of the *convection heat* in the fluid $CV$, term ①, and *transversal conduction heat* transmitted to the fluid $CV$ by the wall, term ②. In order to evaluate the scale of thickness $\delta_T$, we need to work out the following 3 cases, comparing this thickness with the velocity boundary layer thickness $\delta$:

1. First, let us assume that $\delta_T$ layer is "thick" relative to the velocity boundary layer thickness $\delta$ measured at the same $x$, as in Fig. 4.16:

$$\delta_T \gg \delta \qquad\qquad (4.79)$$

Based on scale analysis similitudes, the Equation of energy conservation Eq. (4.78) is rewritten as

$$u\frac{\Delta T}{x} + v\frac{\Delta T}{\delta_T} \sim \alpha_f\frac{\Delta T}{\delta_T^2} \qquad (4.80)$$

According to

$$\frac{\mathbf{w}_\infty}{x} \sim \frac{v}{\delta} \qquad (3.115)$$

the $v$-component scale outside the thin velocity boundary layer (and inside the thermal boundary layer) is $v \sim \mathbf{w}_\infty\delta/x$. This means that the second term on the left-hand side of Eq. (4.80) is

$$v\frac{\Delta T}{\delta_T} \sim \mathbf{w}_\infty\frac{\Delta T}{x}\frac{\delta}{\delta_T} \qquad (4.81)$$

in which $\delta/\delta_T \ll 1$. In conclusion, the *governing Equation for the convection–conduction balance* reduces to

$$u\frac{\Delta T}{x} \sim \alpha_f\frac{\Delta T}{\delta_T^2} \qquad (4.82)$$

and we are left we the determination of the scale of the $u$-component inside the $\delta_T$-thick boundary layer. In this case, inspection of Fig. 4.16 reveals that $u \sim \mathbf{w}_\infty$. Then, we can substitute this estimate into Eq. (4.82) and recall the definition of Peclet number

$$\mathrm{Pe}_L \equiv \frac{\mathbf{w}_\infty L}{\alpha_f} \qquad (4.52)$$

to obtain the first unknown scale

$$\boxed{\delta_T(x) \sim x\mathrm{Pe}_x^{-1/2}} \qquad (4.83)$$

where $\mathrm{Pe}_x$ is a *local Peclet number* based on a downstream position $x$:

$$\mathrm{Pe}_x \equiv \frac{\mathbf{w}_\infty x}{\alpha_f} \qquad (4.84)$$

**The thickness $\delta_T$ is directly proportional to surface length $x$ and inversely proportional to the square root of the Peclet number** based on $x$, $\mathrm{Pe}_x$.
The notion of local Pe is at stake when **evaluating the scale of the wall heat flux**. It is useful to resort to the concept of *local Nusselt number*, starting from the definition of the average Nusselt number

$$|\overline{Nu}_L| \equiv \frac{\overline{h}_T L}{\lambda_f} \tag{4.56a}$$

so we have

$$|Nu_x| \equiv \frac{\dot{q}_{hx} x}{\Delta T \lambda_f} \equiv \frac{h_T x}{\lambda_f} \tag{4.85}$$

where $h_T$ is the *local heat transfer coefficient* given by a definition equivalent to Eq. (4.5):[22]

$$\boxed{h_T \equiv \dot{q}_{hx}/\Delta T} \tag{4.86}$$

Knowing also that

$$\dot{q}_{hx} \sim \lambda_f \frac{\Delta T}{\delta_T} \tag{4.87}$$

we can combine with Eq. (4.83), obtaining the second unknown scale

$$\boxed{Nu_x \sim Pe_x^{1/2}} \tag{4.88}$$

that is, **the greater the Peclet number, the larger the heat flux exchanged by the wall**.
One last implication of inequality Eq. (4.79): substituting Eq. (4.83) and

$$\delta(x) \sim \sqrt{\frac{\nu x}{\mathbf{w}_\infty}} = x Re_x^{-1/2} \tag{3.120}$$

in left- and right-hand sides of the thermal boundary layer assumption $\delta_T \gg \delta$, we see that this is equivalent to

$$\alpha_f > \nu \tag{4.89}$$

In other words, the thermal boundary layer is thicker than the velocity boundary layer when the fluid properties are such that the Prandtl number

$$Pr \equiv \frac{\nu}{\alpha_f} \tag{4.50 revisited}$$

is

$$Pr \ll 1 \tag{4.90}$$

Examples of low-Prandtl-number fluids are the liquid metals such as mercury and liquid sodium.

---

[22] Based on the definitions of average and local convective coefficients, in analogy with Eq. (4.76) it is $\overline{h}_T = \int_0^L h_T(x)\,dx$.

2. In this second case, the $\delta_T$ layer is "thin" relative to $\delta$ measured at the same $x$, Fig. 4.17:

$$\delta_T \ll \delta \tag{4.91}$$

Now, the scale of the $u$-component is

$$u \sim \frac{\delta_T}{\delta} \mathbf{w}_\infty \tag{4.92}$$

This notation can be plugged into the *governing convection–conduction balance* Eq. (4.82), together with the result for the velocity boundary layer thickness $\delta$ Eq. (3.118), to obtain the first unknown scale

$$\boxed{\delta_T(x) \sim x \mathrm{Pr}^{-1/3} \mathrm{Re}_x^{-1/2}} \tag{4.93}$$

Recalling that the relationship between $\delta_T$ and $\dot{q}_{hx}$ was already found with Eq. (4.87), the second scale is readily obtained with Eq. (4.85):

$$\boxed{\mathrm{Nu}_x \sim \mathrm{Pr}^{1/3} \mathrm{Re}_x^{1/2}} \tag{4.94}$$

that is, **the heat flux exchanged by the wall is again proportional to the Prandtl and Reynolds number,** but with different powers, this time. Moreover, in the same way we proceeded earlier, it can be shown that the thermal boundary layer is thinner than the velocity boundary layer when the fluid properties are such that the Prandtl number is

$$\mathrm{Pr} \gg 1 \tag{4.95}$$

Most nonmetallic liquids such as water and oils have relatively large Prandtl number values.

3. With the last case, $\delta_T = \delta$ with fluids having

$$\mathrm{Pr} \simeq 1 \tag{4.96}$$

This happens with most common gases such as air, steam, $CO_2$. By inspecting Eqs. (4.88, 4.94), we readily see that both predict

$$\boxed{\mathrm{Nu}_x \sim \mathrm{Re}_x^{1/2}} \tag{4.97}$$

From what has been showed so far, for the laminar boundary layers at stake here (in which transport by diffusion is not overshadowed by turbulent mixing), it is reasonable to expect that

$$\frac{\delta}{\delta_T} \simeq \mathrm{Pr}^n \tag{4.98}$$

with $n$ a positive exponent. For most applications, it is reasonable to assume a value of $n = 1/3$ [3].

### 4.5.2  Analytical Solutions

A wealth of thermal flow/wall relationships are found in the convective heat transfer applications, so that we must rely on the formulation of Sect. 4.3 (p. 161) for general analysis. However, a certain number of situations in the laminar regime have been solved nonetheless, few of which are presented in the following.[23]

It is customary to focus on the two external and internal configurations, as shown in Figs. 4.2 and 4.3. In both cases, the two main BC kinds are the **uniform surface temperature** Ⓣ and the **uniform surface heat flux** Ⓗ.

**External flow along the plate**   Let us start with the Ⓣ BC, for which the *local heat transfer* is at stake [4]. The classical analytical solution of the boundary layer energy Equation

$$u\frac{\partial T}{\partial x} + v\frac{\partial T}{\partial y} = \alpha_f \frac{\partial^2 T}{\partial y^2} \tag{4.78}$$

together with the continuity Equation

$$\frac{\partial u}{\partial x} + \frac{\partial v}{\partial y} = 0 \tag{3.57}$$

and the momentum Equation

$$u\frac{\partial u}{\partial x} + v\frac{\partial u}{\partial y} = \nu \frac{\partial^2 u}{\partial y^2} \tag{3.114}$$

can be based on the reduction of the PDEs problem to an ordinary differential problem by using a *similarity transformation*, just as alluded to earlier in Sect. 3.5.1 (p. 110).[24] Without digging in the demonstration, which is outside the scope of this book, we report the solution of the scales Eqs. (4.88, 4.94), for two usual ranges of the Prandtl number and the local Nusselt number $\mathrm{Nu}_x$ defined by

$$\mathrm{Nu}_x \equiv \frac{h_T x}{\lambda_f} \tag{4.85 revisited}$$

---

[23] All of these cases pertain to the fluid side, only, that will be referred to later in Sect. 4.5.4 (p. 190) as "segregated solutions," so they imply a conventionally positive Nusselt number, the direction of the heat flux to be determined based on general consideration for each configuration at stake.

[24] As proposed by German mathematician E. POHLHAUSEN in the early twentieth century.

$$\text{(T)} \quad \text{Nu}_x \approx 0.564\text{Pe}_x^{1/2} \quad \text{for} \quad \text{Pr} \lesssim 0.5 \qquad (4.99a)$$

$$\text{(T)} \quad \text{Nu}_x \approx 0.332\text{Pr}^{1/3}\text{Re}_x^{1/2} \quad \text{for} \quad \text{Pr} \gtrsim 0.5 \qquad (4.99b)$$

The corresponding formulas for the average Nusselt numbers $\text{Nu}_L$ defined by

$$\overline{\text{Nu}_L} \equiv \frac{\bar{h}_T L}{\lambda_f} \qquad (4.56a \text{ revisited})$$

follow

$$\text{(T)} \quad \overline{\text{Nu}_L} = 1.13\text{Pe}_x^{1/2} \quad \text{for} \quad \text{Pr} \lesssim 0.5 \qquad (4.100a)$$

$$\text{(T)} \quad \overline{\text{Nu}_L} = 0.664\text{Pr}^{1/3}\text{Re}_x^{1/2} \quad \text{for} \quad \text{Pr} \gtrsim 0.5 \qquad (4.100b)$$

It is worth noting that the former coefficients in Eqs. (4.100) are exactly twice as large as those in the corresponding Eqs. (4.99).

An average Nusselt number formula covering the entire Prandtl number range is suggested [1]:

$$\text{(T)} \quad \overline{\text{Nu}_L} = \frac{0.928\text{Pr}^{1/3}\text{Re}_x^{1/2}}{\left[1 + (0.0207/\text{Pr})^{2/3}\right]^{1/4}} \quad \text{for} \quad \text{Pe} \gtrsim 100 \qquad (4.101)$$

The other basic BC kind is the $\text{(H)}$ one, for which the *local surface free-stream temperature difference* is a stake [1]. The classical analytical solution of the boundary layer energy Equations (4.78, 3.57, 3.114) system can be based on the *integral method* and can be applied to gases and nonmetallic liquids ($\text{Pr} \gtrsim 0.5$). Without digging in the demonstration, we have

$$\text{(H)} \quad T_p(x) - T_\infty \approx \frac{\dot{q}_h x}{0.453\lambda_f\text{Pr}^{1/3}\text{Re}_x^{1/2}} \quad \text{for} \quad \text{Pr} \gtrsim 0.5 \qquad (4.102)$$

In all these cases, we had a developing thermal boundary layer with constant free-stream temperature. Other combination of surface heating conditions can be found, with their solution, in the dedicated literature [1].

The analytical results for wall friction and heat transfer summarized so far are based on the constant-property system Eqs. (3.57, 3.114, 4.78). In reality, rheological properties are not uniform along the heat transfer process, as they depend on the local temperature in the boundary layer flow. It has been verified [1] that the constant-property solving system is accurate enough in evaluating convective heat transfer, provided that the maximum temperature variation experienced by the fluid $(T_p - T_\infty)$ be small relative to its absolute temperature level, $T_p$ or $T_\infty$ (K). In such cases,

**Table 4.2** Average Nusselt number $\overline{Nu}_{D_H}$ defined by above revisited Eq. (4.56), for ⓉＴ and Ⓗ BCs and for some cross-sectional shape [4]

| Cross-sectional shape | Ⓣ $\overline{Nu}_{D_H}$ | Ⓗ $\overline{Nu}_{D_H}$ |
|---|---|---|
| Round | 3.653 | 4.364 |
| Equilateral triangle | 2.470 | 3.111 |
| Square | 2.976 | 3.608 |
| Rectangular, long side $4L$, short side $L$ | 3.391 | 4.123 |
| Wide parallel plates | 7.541 | 8.235 |

the rheological properties needed to calculate the variety of dimensionless numbers can be evaluated at the average temperature of the fluid in the film $\langle T \rangle_f$, or **film temperature**

$$\langle T \rangle_f \equiv \frac{1}{2} \left( T_p + T_\infty \right) \tag{4.103}$$

As we will be seeing, correlations exist for which the temperature variation is duly taken into account.

**Internal flow in the tube**    For both BC kinds Ⓣ Ⓗ, we report in Table 4.2 the solution for the average Nusselt number $\overline{Nu}$ defined by

$$\overline{Nu}_{D_H} \equiv \frac{\overline{h}_T D_H}{\lambda_f} \tag{4.56a revisited}$$

for fully developed flows and the geometries anticipated in Table 3.3. The properties are uniform and computed at the uniform $T_b$ via

$$T_b \equiv \frac{1}{\mathbf{w}_\infty \Omega} \int_\Omega u T \, d\Omega \tag{4.3 revisited}$$

### 4.5.3  Convection Analysis Using Correlations

Beside the numerical solution of governing PDEs, in many cases of interest local and average convective heat transfer can be evaluated starting from appropriate equations that are cast in a general form which is convenient to use for any fluid dynamic regime. This topic is very extended due to the variety of possible flow/surface geometric relationships, and a wealth of configurations is being studied by the year, whose correlations are available on dedicated literature and manuals. Therefore, only limited coverage is made here on this topic, directing the interested readership to some works of broader scope [1, 3–5].

Correlations are useful tools: measurements can be designed, and evaluations can be made when the dimensionless groups the Nusselt number depends on are known, so that the equation's independent variables are varied, by specifying the appropriate coefficient values. Note 22 applies. Unless stated differently, rheological properties are to be evaluated at $\langle T \rangle_f$ for external flows and at $\langle T \rangle_b$ for internal flows. In any case, the resulting values of $\bar{h}_T$ should not vary too much from the indicative values proposed in Table 4.1.

All correlations have limitations which must be noted before they are applied: the geometry configuration, operating and flow regimes, for example, and the applicable range of variables for which the correlation is valid based on the availability of data and/or the extent to which the equation correlated the data. It is important to recall that these *empiricisms* do not always provide very accurate predictions, so errors as high as 25% are not uncommon.

### 4.5.3.1    External Forced Convection

**Plate at uniform conditions**    With reference to the situation in Fig. 3.20, we have seen that for distances $x$ shorter than the critical distance $x'$, for some limited values of the Prandtl number

$$\mathrm{Pr} \equiv \frac{\nu}{\alpha_f} \tag{4.50}$$

the local heat transfer could be evaluated analytically in Eqs. (4.100) or (4.103), for Ⓣ or Ⓗ, respectively. Past $x'$ (in the turbulent regime), the following correlation can be used in a wide range of Pr and for both kinds of BCs [4]:

$$\text{Ⓣ Ⓗ} \qquad \mathrm{Nu}_x = 0.030 \mathrm{Re}_x^{4/5} \mathrm{Pr}^{1/3} \tag{4.104}$$

**External flow normal to a cylinder**    As anticipated in Sect. 3.5.3 (p. 118), the flow conditions in this case are such that the heat transfer will vary in a complicated way around the surface. In this case, depending on the Peclet number

$$\mathrm{Pe}_L \equiv \frac{\mathbf{w}_\infty L}{\alpha_f} \tag{4.52}$$

provided that $\mathrm{Pe}_D = \mathrm{Re}_D \mathrm{Pr} > 0.2$ and for both kinds of BCs, the following turbulent correlation[25] can be used:

$$\text{Ⓣ Ⓗ} \qquad \overline{\mathrm{Nu}}_D = 0.3 + \frac{0.62 \mathrm{Re}_D^{1/2} \mathrm{Pr}^{1/3}}{\left[1 + \left(\frac{0.4}{\mathrm{Pr}}\right)^{2/3}\right]^{1/4}} \left[1 + \left(\frac{\mathrm{Re}_D}{282,000}\right)^{5/8}\right]^{4/5} \tag{4.105}$$

---

[25] As proposed by Churchill-Bernstein [1].

**External flow normal to a sphere**   Provided that $0.71 < Pr < 380$ and $Re_D$ up to $7.6 \times 10^4$, with properties evaluated at $T_p$, the following correlation[26] can be used:

$$\text{Ⓣ} \quad \overline{Nu}_D = 2 + \left[ 0.4Re_D^{1/2} + 0.06Re_D^{2/3} \right] Pr^{0.4} \left( \frac{\mu}{\mu_p} \right)^{1/4} \tag{4.106}$$

### 4.5.3.2 Internal Forced Convection

**Fully developed velocity and temperature in tubes**   With reference to the situation in Fig. 3.19, the most accurate correlation[27] is valid up to $Re_D = 5 \times 10^6$ and for $Pr > 0.5$:

$$\text{Ⓣ Ⓗ} \quad \overline{Nu}_D = \frac{(f_D/8)\,(Re_D - 1,000)\,Pr}{1 + 12.7\,(f_D/8)^{1/2}\,(Pr^{2/3} - 1)} \left[ 1 + \left( \frac{D}{L} \right)^{2/3} \right] \tag{4.107}$$

whereas $f_D$ is the Darcy friction factor seen earlier in Note 17 of Chap. 3 and Fig. 3.7. The factor $D/L$ accounts for entrance effect, while for fully developed flow one sets $D/L = 0$.

### 4.5.3.3 External Natural Convection

**Vertical plate: uniform surface temperature**   With reference to the situation in Fig. 4.15, for any value of the Prandtl number and any fluid regime, depending on the Rayleigh number

$$Ra_L \equiv \frac{g\beta\,(T_p - T_\infty)\,L^3}{\nu\alpha_f} \tag{4.74}$$

provided that $10^{-1} < Ra_L < 10^{12}$, the average heat transfer can be evaluated by the following correlation[28]

$$\text{Ⓣ} \quad \overline{Nu}_L = \left\{ 0.825 + \frac{0.387Ra_L^{1/6}}{\left[ 1 + (0.492/Pr)^{9/16} \right]^{8/27}} \right\}^2 \tag{4.108}$$

---

[26] As proposed by Whitaker [1].

[27] As proposed by Gnielinski. Colburn's, Dittus-Boelter's (accounting for either heating or cooling) and Sieder-Tate's (accounting for temperature variation influence on properties) formulas have also been used with success [1].

[28] As proposed by Churchill-Chu [4].

**Vertical wall: laminar flow, uniform surface heat flux**   Of interest in this case is the determination of the local surface temperature $T_p$ which varies along $y$ (Fig. 4.15). In the laminar regime ($10^4 < \mathrm{Ra}_L < 10^9$) and for any value of the Prandtl number, the local heat transfer is correlated by [4]:

$$ \textcircled{H} \quad \mathrm{Nu}_y \equiv \frac{h_T y}{\lambda_f} = \left[ \frac{\mathrm{Pr}}{4 + 9\mathrm{Pr}^{1/2} + 10\mathrm{Pr}} \mathrm{Ra}_y \right]^{1/4} \tag{4.109} $$

where

$$ h_T = \frac{\dot{q}_{hy}}{T_p(y) - T_\infty} \tag{4.85 revisited} $$

Substituting last Equation, together with the definition of local Rayleigh number

$$ \mathrm{Ra}_y \equiv \frac{g\beta \left[ T_p(y) - T_\infty \right] y^3}{\nu \alpha_f} \tag{4.74 revisited} $$

into Eq. (4.109), we get

$$ \textcircled{H} \quad T_p(y) - T_\infty = \left[ \frac{4 + 9\mathrm{Pr}^{1/2} + 10\mathrm{Pr}}{\mathrm{Pr}} \left( \frac{\nu \alpha_f}{\beta g} \right) \left( \frac{\dot{q}_{hy}}{\lambda} \right)^4 y \right]^{1/5} \tag{4.110} $$

**Inclined walls: laminar flow, uniform surface temperature**   With reference to the situation in Fig. 4.18, it is found in the laminar boundary layer analysis that the momentum Equation

$$ u\frac{\partial v}{\partial x} + v\frac{\partial v}{\partial y} = \nu \frac{\partial^2 v}{\partial x^2} \underbrace{+ g\beta(T - T_\infty)}_{\text{③}} \tag{4.66} $$

holds, except for $g \cos\phi$ (oriented parallel to the wall, in each configuration) replacing $g$ for any value in the source/sink due to the thermal buoyancy force, term ③. For any Prandtl number, in the $-60° < \phi < -60°$ range the average heat transfer can be evaluated by the following correlation[29]

$$ \textcircled{T} \quad \overline{\mathrm{Nu}}_L = 0.68 + \frac{0.67\mathrm{Ra}_L^{1/4} \cos\phi}{\left[ 1 + (0.492/\mathrm{Pr})^{9/16} \right]^{4/9}} \tag{4.111} $$

**Horizontal walls: laminar flow, uniform surface temperature**   The recommended correlations, for a Prandtl number greater than 0.5, are the following [1]:

---

[29] As proposed by Churchill-Chu [1].

**Fig. 4.18** Natural
convection relative to
inclined walls. The arrows
indicate bulk flow. The
nomenclature of the inclined
component of gravity vector
$g \cos \beta$ is also reported. All
possible configurations are
illustrated, from (a) to case
(d), with the inherent average
thermal flux $\dot{q}_h$ exchanged in
each case

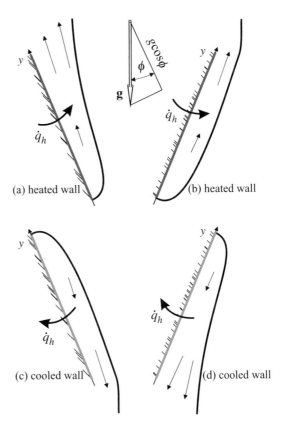

- Heated upper or cooled lower surface

$$\text{(T)} \, \text{(H)} \quad \overline{\text{Nu}}_L = 0.54 \, \text{Ra}_L^{1/4} \quad \text{for} \;\; 2 \times 10^4 < \text{Ra}_L < 8 \times 10^6 \qquad (4.112)$$

$$\text{(T)} \, \text{(H)} \quad \overline{\text{Nu}}_L = 0.15 \, \text{Ra}_L^{1/3} \quad \text{for} \;\; 8 \times 10^6 < \text{Ra}_L < 1.6 \times 10^9 \qquad (4.113)$$

- Heated lower or cooled upper surface

$$\text{(T)} \, \text{(H)} \quad \overline{\text{Nu}}_L = 0.27 \, \text{Ra}_L^{1/4} \quad \text{for} \;\; 10^5 < \text{Ra}_L < 10^{10} \qquad (4.114)$$

In this case, all dimensionless numbers are based on a characteristic length $L$ taken
as the ratio of the heat exchange area of the wall $A$ and $P$: $L = A/P$.

**Vertical cylinder wall: laminar flow, uniform surface temperature**  When the
thermal boundary layer thickness $\delta_T$, in this situation, is much smaller than
its diameter $D$, its curvature does affect the heat exchange and we can employ

Eq. 4.108. This criterion is met when $D/H > \text{Ra}_H^{1/4}$, with $H$ the cylinder height. Otherwise, in the small diameter cylinder limit, the following correlation can be employed [1]:

$$\text{(T)} \quad \overline{\text{Nu}}_H = \frac{4}{3} \left[ \frac{7\text{Ra}_H \text{Pr}}{5\,(20+21\text{Pr})} \right]^{1/4} + \frac{4\,(272+315\text{Pr})\,H}{35\,(64+63\text{Pr})\,D} \tag{4.115}$$

**Horizontal cylinder: laminar flow, uniform surface temperature**    In this case, the correlation resembles the one adopted for a vertical plate Eq. (4.108), as the boundary layer develops similarly [1]:

$$\text{(T)} \quad \overline{\text{Nu}}_D = \left\{ 0.6 + \frac{0.387\,\text{Ra}_D^{1/6}}{\left[ 1 + (0.559/\text{Pr})^{9/16} \right]^{8/27}} \right\}^2 \tag{4.116}$$

**Sphere: laminar flow, uniform surface temperature**    When $\text{Pr} \gtrsim 0.7$ and $\text{Ra}_D < 10^{11}$, the following correlation is appropriate [1]:

$$\text{(T)} \quad \overline{\text{Nu}}_D = 2 + \frac{0.589\,\text{Ra}_D^{1/4}}{\left[ 1 + (0.469/\text{Pr})^{9/16} \right]^{4/9}} \tag{4.117}$$

### 4.5.4 Segregated Versus Conjugate Convection Analysis

Earlier on, we have seen that the experimental determination of the convective heat transfer is based on Newton's Law of convection cooling through the measurement of a suitable temperature difference $\Delta T$

$$\frac{\dot{Q}_h}{A} \equiv \overline{h}_T \Delta T \tag{4.5}$$

based on the configuration at hand, as specified by the definitions of the average heat transfer coefficient $\overline{h}_T$ based on the associated Nusselt number

$$\overline{\text{Nu}}_L \equiv \frac{\overline{h}_T L}{\lambda_f} \tag{4.56a}$$

(or the like, in case of local heat transfer). Indeed, as presented above, a precise computation of the Nusselt number is often prohibitive due to the effects of arbitrary geometries, properties variability, intertwined heat and flow patterns, and turbulence. Even in case of a configuration being adequately covered by one of the above correlations, the heat flow effects that take place in the underlying solid are completely neglected. For example, phase-change and/or conduction temperature alterations may, in reality, affect the pure-convection heat exchange pattern. For these reasons, the results obtained from the application of analytical solutions or correlations are often too simplistic.

As an example, let us consider the common experimental method to calculate the convective heat transfer by providing Joule heating to a solid surface by means of an electric resistor (recall Notes 16 and 36 of Chap. 2): the measurement of current $I$ and voltage drop $\Delta V$ leads directly to $\dot{q}_h$, thanks to the energy conservation applied at the surface. Multiresistors and circuits can be arranged to provide a prescribed surface flux or temperature. Contact or radiative thermometries can even report on surface temperature. However, relying on a convective heat transfer coefficient is equivalent to detach the problem of the heat transfer to/from a surface, from the determination of the temperature distribution underneath, or within the body enveloped by that surface.

In this book, this approach is referred to as the **segregated convection analysis**. As an example, the convection heat transfer problem of determining the transmitted thermal power with the segregated approach would be solved by evaluating $\dot{q}_h$, analytically or empirically, in the $CV$ consisting of the sole fluid phase (Fig. 4.19). Although still popular nowadays in the practice, due to the above drawbacks we emphasize here that this approach often leads to inaccurate or even erroneous results in the presence of said interactions.

A modern alternative to this approach is computational heat transfer. In particular, a **conjugate convection analysis** can be employed using computational tools to determine the temperature distribution simultaneously in both solid and fluid phases or subdomains, as suggested in Fig. 4.20, in order to determine the solution for temperature with no need for empirical assumptions at phase interface: Dirichlet or Neumann BCs assumptions, see Sects. 2.3.1 (p. 26) and 3.3.4 (p. 94), are no longer needed. Now, the subject $CV$ consists in the union of $CV_f$ and $CV_s$. In fact, the continuity of the temperature and heat flux are used as the conjugate boundary conditions to couple the energy equation for both phases, across their interface. In other words, with a conjugated thermal convection analysis *the solution for temperature is shared seamlessly through the interface.* Even the distinction between solid temperature $T_s$ and fluid temperature $T_f$ can be dropped, as we anticipated at the very beginning of this chapter (p. 143). In this case, we can speak of multiphase heat transfer modeling, in one of its simplest cases: in both a stationary solid and its interacting fluid phase.

**Fig. 4.19** Determination of $\dot{q}_h$ by the *segregated* approach, for a fluid $CV$ interacting with a surface

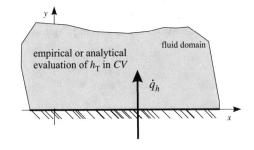

**Fig. 4.20** Determination of $\dot{q}_h$ by the *conjugated* approach for a two-phase solid/fluid $CV$

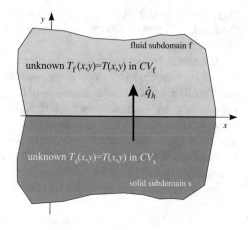

Again in this case, the CFD can be a strategic tool in modeling and computation. Care will be needed to define an appropriate *gridding strategy*, as anticipated in Sect. 2.5.5 (p. 64), along and across the phase interface, and the appropriate discrimination among phase properties, as the same governing equations will be enforced for all of pertinent phases with this approach. For example, in this way the convection heat transfer will be fully described by

$$\overline{\mathrm{Nu}}_L = -\frac{\partial T^*}{\partial y^*} \qquad\qquad \text{(4.56b revisited)}$$

or its local counterpart, and even the direction of heat transfer will be inherently identified by the computation of the derivative, as it will be stretched on the grid points across this interface.

## 4.6   Numerical Solution of Convection of Heat and Other Scalars

The variability and number of analytical results, the validity exceptions on which we just briefly emphasized, are convincing arguments to dig into feasible numerical solution of heat transfer by convection.

Let us focus on the governing Equation for thermal convection

$$\rho_f c_{pf}\left(\frac{\partial T}{\partial \theta} + (\mathbf{w}\cdot\nabla T)\right) = \lambda_f \nabla^2 T \qquad\qquad \text{(4.43)}$$

or similar ones in other coordinate systems, for the sake of simplicity: the system formed with its Navier–Stokes Equations counterpart

**Fig. 4.21** $W$ and $E$ points surrounding the $P$, for the 1-D problem, and the $CV$ faces locations $w$ and $e$

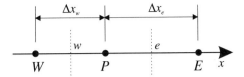

$$\frac{\partial \rho_f}{\partial \theta} + \nabla \cdot (\rho_f \mathbf{w}) = 0 \qquad (3.70a)$$

$$\rho_f \left[ \frac{\partial \mathbf{w}}{\partial \theta} + (\mathbf{w} \cdot \nabla) \mathbf{w} \right] = \mu \nabla^2 \mathbf{w} - \nabla p \pm \mathbf{S} \qquad (3.70b)$$

(or similar ones in other coordinate systems) must be solved to determine the distribution of temperature: when the solution for velocity has been already acquired, or in a simultaneous fashion, as alluded to earlier.

In addition to the thermal problem with macroscopic transport, we also must recall, as alluded to earlier, that the analytical problem of turbulent flow introduced in Sect. 3.3.6 (p. 98) implies the definition of ancillary scalar variables (turbulent kinetic energy, $k$, and turbulent dissipation, $\varepsilon$, in the simplest turbulent modeling) that must be solved numerically. The procedure reviewed already in Sect. 3.6 (p. 120) on numerical solutions of conduction and fluid flow can be then carried on and applied to the *macroscopic transport of heat and any scalar of interest*, even in the presence of turbulence.

We can benefit from having acquired the solutions of Equations such as Eqs. (3.70), so that nothing new still needs to be supplemented. Using again the FV method, and following the same strategy laid out in Sects. 2.5 (p. 55) and 3.6, let us now turn to a method to obtain the numerical (approximate) solution of these Equations, with a special look to the discretization approach and the iterative nature of the adopted method. The procedure will again allow for the transformation of the *differential* equation into a system of *algebraic* equations.

### 4.6.1 Steady-State Heat Convection

We will start with the geometry in Fig. 4.21 enforcing again the FV within the $CV$ along $x$, to identify a number of grid points and their interface points. For each point, an algebraic equation will be written containing the value of temperatures for the cluster of points surrounding the subject grid point. Recalling again the Fundamental Equation of thermal convection in 1-D, in its *variable properties and steady state, with no source and work conversion terms* version Eq. (4.39):

$$\frac{d}{dx}(\rho_f u T) = \frac{d}{dx}\left(\frac{\lambda_f}{c_{pf}}\frac{dT}{dx}\right) \qquad (4.118)$$

These terms are first derivatives in the space domain. Integrating Eq. (4.118) over the $CV$ of $P$ in Fig. 4.21, we get

$$(\rho_f u T)_e - (\rho_f u T)_w = \left.\frac{\lambda_f}{c_{pf}}\frac{dT}{dx}\right|_e - \left.\frac{\lambda_f}{c_{pf}}\frac{dT}{dx}\right|_w \tag{4.119}$$

To simplify the evaluation of the variable at the interface points, similarly than with Eq. (3.119), let us assume that $w$ and $e$ lay midway between neighboring grid points:

$$T_e = \frac{1}{2}(T_E + T_P) \quad \text{and} \quad T_w = \frac{1}{2}(T_P + T_W) \tag{4.120}$$

then we can write, using again the definition of Eq. (2.106):

$$\frac{1}{2}(\rho_f u)_e(T_E + T_P) - \frac{1}{2}(\rho_f u)_w(T_P + T_W) = \left.\frac{\lambda_f}{c_{pf}}\right|_e \frac{T_E - T_P}{\Delta x_e} - \left.\frac{\lambda_f}{c_{pf}}\right|_w \frac{T_P - T_W}{\Delta x_w} \tag{4.121}$$

In order to employ the matrix representation already envisaged in Sects. 2.5 (p. 55) and 3.6 (p. 120), it is useful to cast Eq. (4.121) again in the following form:

$$\boxed{a_P T_P + a_E T_E + a_W T_W = 0} \tag{4.122}$$

where

$$a_E = \frac{1}{2}(\rho_f u)_e - \frac{1}{\Delta x_e}\left.\frac{\lambda_f}{c_{pf}}\right|_e \tag{4.123a}$$

$$a_W = -\frac{1}{2}(\rho_f u)_w - \frac{1}{\Delta x_w}\left.\frac{\lambda_f}{c_{pf}}\right|_w \tag{4.123b}$$

$$a_P = \frac{1}{2}(\rho_f u)_e - \frac{1}{2}(\rho_f u)_w + \frac{1}{\Delta x_e}\left.\frac{\lambda_f}{c_{pf}}\right|_e + \frac{1}{\Delta x_w}\left.\frac{\lambda_f}{c_{pf}}\right|_w$$

$$= -(a_E + a_W) \tag{4.123c}$$

In particular, Eq. (4.123c) has been worked out by recalling Note 56 of Chap. 3. While forming the final algebraic system, a general form like Eq. (4.122) can be cast for any internal grid point. For the Equations involving the boundary grid points, the considerations reported in Sect. 2.5.2.2 (p. 58) apply, along these made in Sects. 2.5.2.3 and 2.5.2.2 for the solution of the resulting algebraic system and the related treatment of nonlinearities.

Discretization of the Fundamental Equation of thermal convection in 2-D is straightforward, with the ideas and principles that were already presented in Sect. 3.6.2 (p. 125).

## 4.6.2 Unsteady-State Heat Convection

Now, let us turn the unsteady-state version of the fundamental Equation of thermal convection in 1-D, by including the transient convective flux $\partial T/\partial\theta$ of Eq. (4.43). Following the same line of thought leading to Eq. (2.126), starting from the grid point values of $T^0$ at time $\theta$, we will evaluate the "new" values $T$ at the time $\theta + \Delta\theta$. Therefore, integrating Eq. (4.43) over the $CV$ of $P$ (Fig. 4.18), we get

$$\int_w^e \int_\theta^{\theta+\Delta\theta} \rho_f \left( \frac{\partial T}{\partial\theta} + u\frac{\partial T}{\partial x} \right) dx d\theta = \int_\theta^{\theta+\Delta\theta} \left[ \left( \frac{\lambda_f}{c_{pf}} \frac{\partial T}{\partial x} \right)_e - \left( \frac{\lambda_f}{c_{pf}} \frac{\partial T}{\partial x} \right)_w \right] d\theta$$
(4.124)

Nothing new needs to be added, with respect to the considerations exploited in Sect. 2.5.3 (p. 61), so that Eq. (4.124) becomes

$$\left[ \frac{1}{2} (\rho_f u)_e (T_E + T_P) - \frac{1}{2} (\rho_f u)_w (T_P + T_W) \right] \Delta\theta + \rho_f \Delta x_{CV} (T_P - T_P^0) = \\ \left( \frac{\lambda_f}{c_{pf}} \bigg|_e \frac{T_E - T_P}{\Delta x_e} - \frac{\lambda_f}{c_{pf}} \bigg|_w \frac{T_P - T_W}{\Delta x_w} \right) \Delta\theta$$
(4.125)

By using again a fully implicit scheme, we come up with

$$\boxed{a_P T_P + a_E T_E + a_W T_W = b}$$
(4.126)

where

$$a_E = \frac{1}{2}(\rho_f u)_e - \frac{1}{\Delta x_e} \frac{\lambda_f}{c_{pf}} \bigg|_e$$
(4.127a)

$$a_W = -\frac{1}{2}(\rho_f u)_w - \frac{1}{\Delta x_w} \frac{\lambda_f}{c_{pf}} \bigg|_w$$
(4.127b)

$$a_P^0 = -\frac{\rho_f \Delta x_{CV}}{\Delta\theta}$$
(4.127c)

$$a_P = -\left( a_E + a_W + a_P^0 \right)$$
(4.127d)

$$b = a_P^0 T_P^0$$
(4.127e)

In case of variable properties, the coefficients are modified as specified earlier. Aspects of the procedure adopted for the steady-state convection, such that boundary conditions and system solution, are applicable to the unsteady situation. As for the temperature-dependent conductivity and source/sink term inclusion case, the iterative procedure seen above also applies, by using the new values of $T_P$.

### 4.6.3 A Heat/Momentum Transfer-Joint Turbulent Model

As alluded to earlier, the procedure leading to Eqs. (4.122) can be applied to numerically solve the joint turbulent transport of heat and momentum. To this end, we need to recall the following:

1. the RANS Equations

$$\rho_f \frac{\partial \overline{w}_i}{\partial \theta} + \rho_f \frac{\partial}{\partial x_j} \left( \overline{w}_i \overline{w}_j \right) = \frac{\partial}{\partial x_j} \left[ (\mu + \mu_t) \frac{\partial \overline{w}_i}{\partial x_j} \right] - \frac{\partial \overline{p}}{\partial x_i} \qquad (3.102)$$

   to solve for the turbulent velocity $\overline{w}$ and pressure $\overline{p}$, when a closure turbulence model is adopted for the turbulence eddy viscosity $\mu_t$;
2. a suitable turbulence model such as

$$\frac{\partial (\rho_f k)}{\partial \theta} + \frac{\partial \left( \rho_f \overline{w}_j k \right)}{\partial x_j} = \frac{\partial}{\partial x_j} \left[ \left( \mu + \frac{\mu_t}{\sigma_k} \right) \frac{\partial k}{\partial x_j} \right] - \rho_f \varepsilon + S_k \qquad (3.96)$$

$$\frac{\partial (\rho_f \varepsilon)}{\partial \theta} + \frac{\partial \left( \rho_f \overline{w}_j \varepsilon \right)}{\partial x_j} = \frac{\partial}{\partial x_j} \left[ \left( \mu + \frac{\mu_t}{\sigma_\varepsilon} \right) \frac{\partial \varepsilon}{\partial x_j} \right] + S_\varepsilon \qquad (3.98)$$

   along with their ancillary definitions, which solves for the turbulent kinetic energy $k$ and the turbulent dissipation $\varepsilon$;
3. the Equation of turbulent thermal convection

$$\frac{\partial \overline{T}}{\partial \theta} + \frac{\partial}{\partial x_j} \left( \overline{T} \overline{w}_j \right) = \frac{\partial}{\partial x_j} \left( (\alpha_f + \varepsilon_\alpha) \frac{\partial \overline{T}}{\partial x_j} \right) \qquad (4.47)$$

   to solve for the turbulent temperature $\overline{T}$.

The RANS Equations set can be adequately attacked, by enforcing the same solving sequence described by Sect. 3.6.4 (p. 133) for velocity and pressure, once the dynamic viscosity $\mu$ is incremented by a turbulent eddy viscosity term $\mu_t$ into the diffusive fluxes of Eqs. (3.148). In a similar way, the steady-state of turbulent thermal convection can enjoy the same development leading to the solving Eq. (4.122), once the thermal diffusivity $\alpha_f$ is incremented by a eddy diffusivity term $\varepsilon_\alpha$.

   As for the solution of governing equation for scalars $k$ and $\varepsilon$, the development leading to Eq. (4.122) can be exploited again, with few adding remarks concerning the diffusive and source terms, as shown in the following.

   We will start again with the geometry in Fig. 4.18 enforcing again the FV within the $CV$ along $x$, to identify a number of grid points and their interface points. For each point an algebraic equation will be written containing the value of turbulent kinetic energy and turbulent dissipation, for the cluster of points surrounding the subject grid point. Recalling $k$ and $\varepsilon$ balances Eqs. (3.96) and (3.98), respectively, in their *steady state* for the sake of simplicity, and using the mean (time-averaged) value of the $u$-component of velocity:

$$\frac{d\,(\rho_f \bar{u} k)}{dx} = \frac{d}{dx}\left[\left(\mu + \frac{\mu_t}{\sigma_k}\right)\frac{dk}{dx}\right] - \rho_f \varepsilon + S_k \tag{4.128}$$

$$\frac{d\,(\rho_f \bar{u} \varepsilon)}{dx} = \frac{d}{dx}\left[\left(\mu + \frac{\mu_t}{\sigma_\varepsilon}\right)\frac{d\varepsilon}{dx}\right] + S_\varepsilon \tag{4.129}$$

then we can write, using again the definition of Eq. (2.106):

$$\frac{1}{2}\,(\rho_f \bar{u})_e\,(k_E + k_P) - \frac{1}{2}\,(\rho_f \bar{u})_w\,(k_P + k_W) =$$
$$\left(\mu + \frac{\mu_t}{\sigma_k}\right)_e \frac{k_E - k_P}{\Delta x_e} - \left(\mu + \frac{\mu_t}{\sigma_k}\right)_w \frac{k_P - k_W}{\Delta x_w} - (\rho_f \varepsilon - S_k)\int_w^e dx \tag{4.130}$$

and

$$\frac{1}{2}\,(\rho_f \bar{u})_e\,(\varepsilon_E + \varepsilon_P) - \frac{1}{2}\,(\rho_f \bar{u})_w\,(\varepsilon_P + \varepsilon_W) =$$
$$\left(\mu + \frac{\mu_t}{\sigma_\varepsilon}\right)_e \frac{\varepsilon_E - \varepsilon_P}{\Delta x_e} - \left(\mu + \frac{\mu_t}{\sigma_\varepsilon}\right)_w \frac{\varepsilon_P - \varepsilon_W}{\Delta x_w} + S_\varepsilon \int_w^e dx \tag{4.131}$$

It is clear that the value of $\varepsilon$ to be taken in the source/sink of Eq. (4.130) is to be considered as constant, while solving for $k$. In order to employ the matrix representation already envisaged in Sects. 2.5 (p. 55) and 3.6 (p. 120), it is useful to cast Eq. (4.130) again in the following form:

$$\boxed{a_P k_P + a_E k_E + a_W k_W = b} \tag{4.132}$$

where

$$a_E = \frac{1}{2}(\rho_f \bar{u})_e - \frac{1}{\Delta x_e}\left(\mu + \frac{\mu_t}{\sigma_k}\right)_e \tag{4.133a}$$

$$a_W = -\frac{1}{2}(\rho_f \bar{u})_w - \frac{1}{\Delta x_w}\left(\mu + \frac{\mu_t}{\sigma_k}\right)_w \tag{4.133b}$$

$$a_P = \frac{1}{2}(\rho_f \bar{u})_e - \frac{1}{2}(\rho_f \bar{u})_w + \frac{1}{\Delta x_e}\left(\mu + \frac{\mu_t}{\sigma_k}\right)_e + \frac{1}{\Delta x_w}\left(\mu + \frac{\mu_t}{\sigma_k}\right)_w$$
$$= -(a_E + a_W) + (\rho_f \bar{u})_e - (\rho_f \bar{u})_w = -(a_E + a_W) \tag{4.133c}$$

$$b = -(\rho_f \varepsilon - S_k)\,\Delta x_{CV} \tag{4.133d}$$

while Eq. (4.131) is cast in the following form:

$$\boxed{a_P \varepsilon_P + a_E \varepsilon_E + a_W \varepsilon_W = b} \tag{4.134}$$

where

$$a_E = \frac{1}{2}(\rho_f \bar{u})_e - \frac{1}{\Delta x_e}\left(\mu + \frac{\mu_t}{\sigma_\varepsilon}\right)_e \tag{4.135a}$$

$$a_W = -\frac{1}{2}(\rho_f \bar{u})_w - \frac{1}{\Delta x_w}\left(\mu + \frac{\mu_t}{\sigma_\varepsilon}\right)_w \tag{4.135b}$$

$$a_P = \frac{1}{2}(\rho_f \bar{u})_e - \frac{1}{2}(\rho_f \bar{u})_w + \frac{1}{\Delta x_e}\left(\mu + \frac{\mu_t}{\sigma_\varepsilon}\right)_e + \frac{1}{\Delta x_w}\left(\mu + \frac{\mu_t}{\sigma_\varepsilon}\right)_w$$
$$= -(a_E + a_W) + (\rho_f \bar{u})_e - (\rho_f \bar{u})_w = -(a_E + a_W) \tag{4.135c}$$
$$b = S_\varepsilon \Delta x_{CV} \tag{4.135d}$$

The constants defined in Eqs. (3.103) apply everywhere in Eqs. (4.132, 4.134).

Now, we realize that a great number of *nonlinearity* issues exists, in the system of Eqs. (4.132, 4.134), for which recursive resort to the procedure explained in Sect. 2.5.2.4 (p. 60) is required:

- $k$ and $\varepsilon$ are present in Eqs. (4.133a–4.133c) and in Eqs. (4.135a–4.135c) as components of the turbulent eddy viscosity $\mu_t$ via Eq. (3.100);
- in Eq. (4.133d), the definition of Eq. (3.97) of the rate of production of turbulent energy by the mean flow, $S_k$, requires the evaluation of the right-hand-side terms of Eq. (3.95), with the following two additional sources of nonlinearities:

  1. the contribution of the other two components of velocity
  2. the insertion of the turbulent kinetic energy $k$ itself

- similarly, in Eq. (4.135d), the definition of Eq. (3.99) of the rate of production of dissipation, $S_\varepsilon$, requires the evaluation of the turbulent dissipation $\varepsilon$ itself.

While forming the final algebraic system, general forms like Eqs. (4.132, 4.134) can be cast for any internal grid point. For the Equations involving the boundary grid points, the considerations reported in Sect. 2.5.2.2 (p. 58) apply. Of course, a similar development can be carried over for the other velocity components.

It is clear that hand-programming of the above procedure becomes soon tedious, but with no further conceptual difficulty. Indeed, the interested reader should be convinced at this point that the solution of thermal flow problems in the presence of turbulence is feasible even with a simplified $k - \varepsilon$ model. In many cases, however, the configuration is such that more complex turbulent modeling is necessary, for which the programming engineer can resort, with appropriate learning and dedication, to CFD software alluded to in Sect. 3.6.6 (p. 137).

## 4.7  Further Reading

- Development of fundamental equations of thermal convection. Bejan, A.: Convection Heat Transfer. Chapters 2 and 4. Wiley, New York (1995)
- Dimensionless numbers in convection heat transfer and their meaning. Kuneš, J.: Dimensionless Physical Quantities in Science and Engineering. Elsevier, London (2012)
- Combined modes and process applications. Jaluria, Y., Torrance, K. E.: Computational Heat Transfer. Hemisphere Publishing, Washington (1986)
- Analytical methods development in convection heat transfer. Kays, W.M., Crawford, M.E.: Convective Heat and Mass Transfer. Chapters 6–13 and 16. McGraw-Hill, New York (1980)
- Dimensionless groups; Extensive analytical methods and correlations in convection heat transfer. Shah, R.K., London, A.L.: Laminar Flow Forced Convection in Ducts. Chapters 3–6. Academic Press, New York (1978)
- Design of heat transfer equipment. Hewitt, G.F., Shires, G.L., Bott, T.R.: Process Heat Transfer. CRC press, Boca Raton (1994)
- Correlation of forced, external convection. Rubesin, M.W, Inouye, M., Parikh, P.G.: Forced Convection, External Flows. In: Rohsenow, W.M., Hartnett, J.P., Cho, Y.I. (Eds.) Handbook of Heat Transfer. McGraw-Hill, New York (1998)
- Correlation of forced, internal convection. Ebadian, M.A., Dong, Z.F.: Forced convection, Internal Flow in Ducts. In: Rohsenow, W.M., Hartnett, J.P., Cho, Y.I. (Eds.) Handbook of Heat Transfer. McGraw-Hill, New York (1998)
- Correlation of natural convection. Raithby, G.D., Hollands, K.G.T.: Natural Convection. In: Rohsenow, W.M., Hartnett, J.P., Cho, Y.I. (Eds.) Handbook of Heat Transfer. McGraw-Hill, New York (1998)

## References

1. Bejan, A.: Heat Transfer. Wiley, New York (1993)
2. Bird, R.B., Stewart, W.E., Lightfoot, E.N.: Transport Phenomena. Wiley, New York (2002)
3. Bergman, T.L., Incropera, F.P., Lavine, A.: Fundamentals of Heat and Mass Transfer. Wiley, New York (2011)
4. Jiji, L.M.: Heat Transfer Essentials: A Textbook. Begell House, New York (1993)
5. Mills, A.F.: Heat Transfer. Richard D. Irwin, Boston (1992)

# Chapter 5
# Mass Transfer by Diffusion and Convection

**Abstract** *Transfer of mass*, in the sense of *chemical or biological species*, is the third and last physical mechanism we encounter: with the analysis of *diffusion* and *mass convection* a preliminary outlook on transfer phenomena is completed. Strong similarities exist between heat and species transport: we will use our acquired knowledge to describe the **species transport in stationary media** and a fluid stream that acts as a **species carrier between two media in relative motion**. After a brief reference to the basic physical mechanism, as we recognize that mass of a given substance is driven by a *concentration difference*, we start by exploiting first the *macroscopic balance* for diffusion in the simplest case of **binary systems**. So far our attention was drawn to the description of temperature and flow fields: now, we derive and integrate *governing differential equations* as usual in various cases, following the *microscopic balance*, but as one substance moves relative to another, we arrive to the **distribution of a mass concentration scalar**. Along the same line that we exploit so far, with few distinctions only, we will find that many concepts have been already laid down, leading to the *numerical solution*, so we will be ready to move to the last chapter to scrutinize some PCB phenomena of interest.

## 5.1 Mass Transfer: The Underlying Physics and Basic Definitions

In the course of the incremental learning which has been proposed in this book with the sequence of transport mechanisms, we proposed to focus, from time to time, to the PCB notations that may characterize the variety of realistic modeling of our products, processes, and plants, some of which will be presented in the last Chap. 6. By acknowledging this, we are faced now with the one mechanism that summarizes, in some way, all other ones but also contains few distinguishing aspects that need to be highlighted.

---

The original version of this chapter was revised: Belated corrections have been incorporated. The correction to this chapter is available at https://doi.org/10.1007/978-3-319-66822-2_7

© Springer International Publishing AG 2018
G. Ruocco, *Introduction to Transport Phenomena Modeling*,
https://doi.org/10.1007/978-3-319-66822-2_5

### 5.1.1 Multiphysics: Phases, Components, Species

The concept of "multiphysics" or PCB phenomena that we are adopted so far encases some specific nomenclature needing proper recall and specification [1]. A *multiphase* system is one characterized by the simultaneous presence of several *phases* (with the implied meaning drawn from *physics*), the two-phase system being a common, relatively simple case. The term "two-component" or "binary" is also sometimes used to describe substance flows consisting in two different chemical/biological substances or *species* (with the implied meaning drawn from *chemistry/biology*). Some binary systems (mostly liquid–liquid) consist of a single phase but may be identified as two-phase, the term "phase" being applied to each of the components.

Multiphase systems, where flow (momentum transfer) is a prevailing aspect, is a vast field of study, as the phases passing through the $CV$ may be solid, liquid, gas or a combination thereof, with many important applications involved. However, since the mathematics describing two-phase and binary flows is the same, the two expressions are often treated as synonymous. These **multiphase flows** can also be classified considering the structure of the interfaces. Phases or components can be *separated* (immiscible), with a well-defined, geometrically simple interface. When the interface is multiple and complex, the phases are *disperse* meaning that the transport properties are those of the bulk flow. A third class may be found when separated and dispersed phases coexist forming a *mixed* phase, often implying a gradual change one into another.

For example, steam–liquid water systems are two-phase, while air–liquid water systems are binary: but in both cases, the interface may depend on mass flow ratio, geometry, and regimes.

Moreover, **mixtures** (when no such prevailing flow aspect exists) can be classified based on their inherent interface, as well. Whether there is none, a *homogeneous* mixture occurs (sometime with physical properties different than those of its individual components). An equivalent term equivalent to "disperse flow" is the *dilute mixture*. Otherwise, we have a *heterogeneous* mixture.

For example, air is a homogeneous gas mixture, common salt or sugar dissolved in liquid water form a homogeneous liquid mixture. In these cases, we have a *solution* mixture, consisting *one or more solutes in a solvent*. The other two cases of mixtures are the *colloid* (when the heterogeneous nature is limited to a microscopical level) and the *suspension*, each one with many applicative cases.[1]

### 5.1.2 Molecular Flow of Mass Components

Let us consider a glass of liquid water. In order to describe the onset of mass diffusion and related concentration distribution in the water, we can imagine a small dye quantity being dropped in the glass as in Fig. 5.1, forming a binary mixture. As soon

---

[1]Mixture applications are summarized in https://en.wikipedia.org/wiki/Mixture.

**Fig. 5.1** A small quantity of a common dye (such as methylene blue) is dropped into a glass of water. In this particular case, over the time, the water gets colored uniformly, due to diffusion of methylene blue into the available $CV$

as the dye makes direct contact with the water, the chemical species flows throughout the available $CV$, eventually coloring the water uniformly after some time $\theta$.

Diffusion deals with the three effects applying to mass that is managed when different substances in chemical or biological disequilibrium are in intimate contact: the *net transfer of mass*, the *mass inertia*, and the *internal mass creation/destruction*. Earlier on, while examining the thermal conduction in Sect. 2.1.1 (p. 13), we found the corresponding terms being invoked for the transport of heat: the first two terms (net transfer of heat and the thermal inertia) being governed by the relation between heat flow and the specific medium, and the third one (the internal heat generation/dissipation) related to the inherent multiphysics. In contrast, in the present case (besides the driving force for the substance flow) *these three effects are found in the exclusive relationships between the species that make up the medium*, including the possible transformation of one species into another one. These relationships are an important difference between molecular heat and mass transfer.

Let us start our analysis by taking a closer (microscopic) look at the given chemical or biological binary mixture in equilibrium.

## 5.1.3 Fick's Law of Diffusion

### 5.1.3.1 Concentration Profile in a Diffusive Medium

Let us consider a $CV$ consisting in a uniform quantity of a species 1 laying in equilibrium in a layer of a species 2 (solid, or fluid at rest) as in Fig. 5.2a. The layer has a finite thickness $L$ along $y$, while it is unconfined along the other two coordinates.

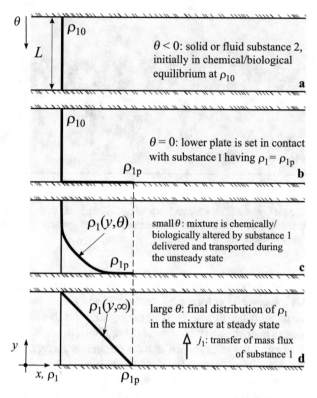

**Fig. 5.2** Variation, during time progress $\theta$ (from top **a**, to bottom **d**), of the concentration profile $\rho_1$ in the substance 2, due to the availability of substance 1 at layer's lower side. With reference to Note 3, in this case we observe a linear profile. The direction of the mass flux of substance 1 $j_1$ is evidenced at lower right

This large width (area), in any $x - z$ plane, will have a size $A$. In general, each species $i$ is quantified by its mass $m_i$ over the available total mixture volume $V$:

$$\rho_i \equiv \frac{m_i}{V} \tag{5.1}$$

with $\rho_i$ the *mass concentration* (with $[\rho_i]$=kg/m$^3$).[2] Let us assume that species 1 is initially at $\rho_{10}$,[3] while at a time $\theta = 0$ an additional amount of substance 1 is made available, at the lower side (or limiting plate) of the layer, with a mass concentration $\rho_{1p}$ (Fig. 5.2b).

A mass concentration profile or distribution of species 1 $\rho_1(y, \theta)$, variable in space and time, is therefore created as in Fig. 5.2c due to its **molecular transport** into and across species 2: after some time (Fig. 5.2d), when the steady state is restored,

---

[2] An alternative definition of mass concentration is *partial density* or *concentration* in short.

[3] The thick line in Fig. 5.2a is normal to the $\rho_1$ axis.

molecules at the lower layer side are found at the same $\rho_{1p}$, while those at the upper layer still stay at the same initial $\rho_{10}$.[4] In other words,

$$\rho_1(0, \infty) = \rho_{1p}, \quad \rho_1(L, \infty) = \rho_{10} \tag{5.2}$$

### 5.1.3.2   Flow of a Substance by Diffusion

Let us denote with $j_1$ the **diffusion mass flux of species 1**, that is, the mass rate $\dot{m}_1$ of species 1 transferred by diffusion along the $y$-direction through the layer surface $A$, applied to the layer's lower side in order to maintain the mass concentration difference or gradient $\Delta\rho_1 = \rho_{1p} - \rho_{10}$ across the layer. For the situation reproduced in Fig. 5.2d, it is therefore:

$$\frac{\dot{m}_1}{A} \equiv j_1 \propto \frac{\Delta\rho_1}{L} \tag{5.3}$$

The flux $j_1$ causes a constant **net transport of species 1 through the adjacent molecules**, *in the direction taken by mass concentration decrease*. Equation (5.3) evidences the existing link between the *mass transferred through the layer* and the *observable concentration field*. In other words, the **concentration gradient is the driving force of mass transfer**. For the mass rate and the mass flux, it is $[\dot{m}]$=kg/s and $[j_1]$=kg/m$^2$s, respectively; while it is $[L]$=m and $[\Delta\rho]$=kg/m$^3$ for the layer thickness and the difference of mass concentration, respectively.

### 5.1.3.3   Link Between Concentration Difference and Mass Transfer

Once a $CS$ and a $CV$ are assigned, the essential problems in studying mass transfer are

- the *determination of the mass rate transported under a given concentration difference and flow field* through the $CS$, and
- the *determination of the concentration distribution* on the $CS$ or in the $CV$.

It is clear that the problem of mass transfer comprises the transport by **molecular mass diffusion** and the transport by **macroscopical mass convection**, in complete analogy with convection heat transfer, Sect. 4.1.2.3 (p. 149).

**Molecular mass diffusion**   The Fourier's Law of conduction linked the thermal conductivity of a stationary medium and the temperature difference with the transferred thermal power:

$$\frac{\dot{Q}_\lambda}{A} = \lambda_s \frac{\Delta T}{L} \tag{2.3 revisited}$$

---

[4]The concentration profile will be linear with $y$ or otherwise, depending on the material characteristic with respect to diffusion.

Similarly, the proportionality parameter, or *material property* inherent to Eq. (5.3), is characteristic of species 1 to diffuse through species 2 in a stationary medium, and it is called its **mass diffusivity** $D_{12}$:

$$j_1 = D_{12}\frac{\Delta\rho_1}{L} \qquad\qquad (5.4)$$

This is the **Fick's Law of mass diffusion**:[5] *the mass flux of species 1 transferred by diffusion through a layer of species 2 is equal to the product of its mass diffusivity, and the mass concentration gradient through the layer.*

In the present framework, a *differential form* (the relationship obtained as the layer thickness $L$ tends to 0) is preferred to the *algebraic form* of Eq. (5.4), except when a *macroscopic balance* is sought (as will be seen later). Having noted that the mass flux is positive when the concentration gradient is negative, we can introduce the **Fick's Law of diffusion in the steady state in differential form**:

$$j_1 = -D_{12}\frac{d\rho_1}{dy} \qquad\qquad (5.5)$$

Equation (5.5) is valid for any binary solid, liquid, or gas solution, provided that the medium is isothermal and isobaric; nevertheless, it is a good engineering approximation in many non-equilibrium systems, where the concentration gradient is superimposed on temperature gradient.

The mass diffusivity $D$, as any other material property, may change in any material point in the $CV$. In this case, Eqs. (5.4, 5.5) refer to a situation with *uniform mass diffusivity*.

**Macroscopical mass convection**     Let us suppose now that species 2 is in motion. We already used the concept of convective thermal flux, that is, the thermal power transferred by convection through the exposed surface, depending on the temperature difference or gradient:

$$\frac{\dot{Q}_h}{A} \equiv \bar{h}_T\Delta T \qquad\qquad (4.5)$$

Similarly, we see that mixture properties and velocity and concentration distributions *when mass is in macroscopical motion* all have a strong influence on the determination of the mass exchange by convection. We will be soon examining in some detail the interaction between fluid dynamics and mass convection, *in the boundary layer* region where its effects are found.

---

[5] As proposed by German physiologist A. E. FICK, in the late nineteenth century.

Therefore, as the concentration field is now intertwined with the distribution of velocity, we can invoke the **average convective mass transfer coefficient** $\bar{h}_M$ (with Note 2 of Chap. 4 in mind)

$$\boxed{j_1 \equiv \bar{h}_M \Delta \rho_1} \tag{5.6}$$

with $[\bar{h}_M]$=m/s. One can evaluate $\bar{h}_M$ by looking at the fluid side of the confining surface in Figs. 4.2 and 4.3, interchanging the distribution of the scalar temperature with the distribution of the scalar mass concentration: due to the no-slip condition at the walls, the *mass flux of species 1 that traverses the laminar sublayer of species 2 is ruled by sole diffusion in the fluid*, whose mass diffusivity thermal conductivity is $D_{12}$. Then, just as we did for the thermal balance

$$-\lambda_f \frac{dT}{dx} = \bar{h}_T \, (T - T_\infty) \tag{2.33b revisited}$$

we can write

$$- D_{12} \frac{d\rho_1}{dy} = \bar{h}_M \, (\rho_1 - \rho_{1\infty}) \tag{5.7}$$

and

$$\bar{h}_M = - \frac{D_{12}}{\rho_1 - \rho_{1\infty}} \frac{d\rho_1}{dy} \tag{5.8}$$

Indeed, the definition of convective mass transfer coefficient may be well used in analogy with Sect. 2.2.3 (p. 23), when the composite resistance and conductance come in handy in mass transfer networks, in simultaneous diffusion and mass convection, to come up with a global exchange coefficient for a composite medium, or **overall mass transfer coefficient** $U_M$. Moreover, all considerations made in Sect. 4.1.2.4 (p. 150) on the operating regimes and bulk flow assumption apply for mass transfer, as well.

With these definitions, the need of the concentration field solution in the fluid in order to evaluate the convective mass flux is ascertained.

### 5.1.4 Mass and Molar Relationships for Mixtures

This transfer mechanism requires a number of definition concerning macroscopic (bulk) or molar mass, as the information on each mixture component is generally needed.

First, the *density* of a $n$-components mixture $\rho$ is given by

$$\rho \equiv \sum_{i=1}^{n} \rho_i \qquad (5.9)$$

In some applications, specially when dealing with gaseous mixtures, the concept of dimensionless *mole fraction* of species containing $N$ moles moles is frequently used:

$$x_i \equiv \frac{N_i}{N} \qquad (5.10)$$

where $N_i$ is the *number of moles* of substance $i$ in the mixture. Clearly, it is

$$\sum_{i=1}^{n} x_i = 1 \qquad (5.11)$$

In case of an ideal gas mixture, mole fractions are proportional to the respective *partial pressures* $p_i$ via the *mixture total pressure* $p$:

$$x_i = \frac{p_i}{p} \qquad (5.12)$$

Mole fraction $x_i$ must be distinguished from the dimensionless *mass fraction* $\omega_i$:

$$\omega_i \equiv \frac{\rho_i}{\rho} \qquad (5.13)$$

The component and mixture number of moles, $N_i$ and $N$, are proportional to the *component* and *mixture molar mass*, $M_i$ and $M$,[6] respectively:

$$N_i \equiv \frac{m_i}{M_i}; \qquad N \equiv \frac{m}{M} \qquad (5.14)$$

with $M \equiv \sum_{i=1}^{n} x_i M_i$. The *molar concentration* $c_i$[7] is also frequently used

$$c_i \equiv \frac{N_i}{V} = \frac{\rho}{M} x_i \qquad (5.15)$$

with $[c_i] = kmol/m^3$. The molar concentration must be distinguished from previously introduced mass concentration $\rho_i$:

$$c_i = \frac{\rho_i}{M_i} \qquad (5.16)$$

---

[6]The molar mass $M_i$ is also called the *molecular weight*, $[M] = kg/kmol$. A related applicative case is presented in Sect. 6.3.

[7]$c_i$ and $\omega_i$ nomenclature is sometime used interchangeably, in the associated literature; therefore, care is needed in managing the related dimensions.

With the definition of Eqs. (5.15), (5.5) can be restated on a molar basis, yielding the *diffusion molar flux* $\hat{j}$ for species 1 (with $[\hat{j}]=$kmol/m$^2$s)

$$\boxed{\hat{j}_1 = -D_{12}\frac{dc_1}{dy}}$$   (5.17)

### 5.1.5 The Driving Material Property

The mass diffusivity $D$ is the property parameter [2, 3] that will be used in the development of the governing equations for the present transfer mechanism. Common substance couplings feature different $D$ values. Let us start with *binary gaseous mixtures*, at atmospheric pressure and for specific temperatures. Table 5.1 shows the value of the *mutual mass diffusivity* $D$ ($[D]=$m$^2$/s)[8] for selected substance couplings, where

$$D \equiv D_{12} = D_{21}$$   (5.18)

In other words, the mass diffusivity of species 1 into 2 equals the diffusivity of species 2 into 1.[9]

The mass diffusivities of *liquid mixtures* are generally $10^4$ to $10^5$ times smaller than the diffusivities seen with the gaseous mixtures. In binary liquid mixtures, the mass diffusivity is a function of composition, unlike in binary gaseous mixtures. This effect becomes negligible when only a small amount of solute is mixed with the solvent, with $x_{max} = 0.05$. Table 5.2 shows the value of diffusivity $D$ for selected solutes in liquid water. Generally, diffusivity in these cases increases with the temperature/ dynamic viscosity ratio of solvent.

---

[8]It is worth to note that mass diffusivity $D$, thermal diffusivity $\alpha$ Eq. (2.9), and kinematic diffusivity $\nu$ Eq. (3.6) all share the same units, indicating once again strong analogies between different transport phenomena.

[9]The occurrence described by Eq. (5.18) is also called the *equimolar counterdiffusion*. This can be demonstrated by noting that, in correspondence to the mass flux of species 1 $j_{1y}$ directed upward, a mass flux of species 2, $j_{2y}$, superimposes downward in any $y$-constant plane (Fig. 5.2). On molar basis, since the mixture is stationary, the net flow rate must be zero:

$$\hat{j}_{1y} + \hat{j}_{2y} = 0$$

Applying Eq. (5.17) one has

$$-D_{12}\frac{dc_1}{dy} = D_{21}\frac{dc_2}{dy}$$

or, with the conversion of Eq. (5.15),

$$-D_{12}\frac{dx_1}{dy} = D_{21}\frac{dx_2}{dy}$$

Hence, knowing that $x_1 = 1 - x_2$ based on Eq. (5.11), we verify the validity of Eq. (5.18).

**Table 5.1** Mass diffusivity $D$ (m$^2$/s) of binary gaseous mixtures at atmospheric pressure and given temperatures $T$ (K)

| Gaseous mixture | $D$ | $T$ |
|---|---|---|
| Air–benzene | $0.77 \times 10^{-5}$ | 273 |
| Air–carbon dioxide | $1.42 \times 10^{-5}$ | 276 |
|  | $1.77 \times 10^{-5}$ | 317 |
| Air–ethanol | $1.45 \times 10^{-5}$ | 313 |
| Air–naphtalene | $5.13 \times 10^{-6}$ | 273 |
| Air–water vapor | $2.60 \times 10^{-5}$ | 298 |
|  | $2.88 \times 10^{-5}$ | 313 |
| Carbon dioxide–oxygen | $1.53 \times 10^{-5}$ | 293 |
| Carbon dioxide–water vapor | $1.98 \times 10^{-5}$ | 307 |
| Hydrogen–water vapor | $9.15 \times 10^{-5}$ | 307 |
| Methane–water vapor | $3.56 \times 10^{-5}$ | 352 |
| Oxygen–water vapor | $3.52 \times 10^{-5}$ | 352 |

**Table 5.2** Mass diffusivity $D$ (m$^2$/s) of gases and organic solutes in dilute acqueous solutions at given temperatures $T$ (K)

| Solute | Solvent | $D$ | $T$ |
|---|---|---|---|
| Acetone | Water | $1.16 \times 10^{-9}$ | 293 |
| Air | Water | $2.50 \times 10^{-9}$ | 293 |
| Benzene | Water | $1.02 \times 10^{-9}$ | 293 |
| Carbon dioxide | Water | $1.92 \times 10^{-9}$ | 298 |
| Chlorine | Water | $1.25 \times 10^{-9}$ | 298 |
| Ethanol | Water | $0.84 \times 10^{-5}$ | 298 |
| Ethylene glycol | Water | $1.04 \times 10^{-9}$ | 293 |
| Nitrogen | Water | $2.60 \times 10^{-9}$ | 293 |
| Oxygen | Water | $2.10 \times 10^{-9}$ | 298 |
| Propane | Water | $0.97 \times 10^{-9}$ | 293 |
| Urea | Water | $1.20 \times 10^{-9}$ | 293 |

The mass diffusivities through *solids* are even smaller than those seen so far. In this case, the *structure* of the solid such as channeling and pores dictates on the diffusion phenomenon. Table 5.3 shows the value of diffusivity $D$ for selected gases through few common gasket and package materials, at room temperature ($T = 298\ K$).

Finally, biostructures can be subject to diffusion phenomena, as well. Table 5.4 reports on few cases drawn on the modeling examples reported on in Chap. 6.

**Table 5.3** Mass diffusivity $D$ ($m^2/s$) of gases through solid materials at ambient temperature

| Solid | $O_2$ | $N_2$ | $CO_2$ | $CH_4$ |
|---|---|---|---|---|
| Natural rubber | $1.58 \times 10^{-10}$ | $1.10 \times 10^{-10}$ | $1.10 \times 10^{-10}$ | $0.89 \times 10^{-10}$ |
| Silicon rubber | $17.0 \times 10^{-10}$ | $13.2 \times 10^{-10}$ | | |
| Polyethylene | $0.17 \times 10^{-10}$ | $0.093 \times 10^{-10}$ | $0.124 \times 10^{-10}$ | $0.057 \times 10^{-10}$ |
| Polystyrene | $0.11 \times 10^{-10}$ | | $0.058 \times 10^{-10}$ | |

**Table 5.4** Indicative values of mass diffusivity $D$ ($10^{-9} m^2/s$) of selected chemical and biological species in biomedia. *Bacteria diffusivity depends on temperature (through activity and metabolism). $^+$Water diffusivity in food media depends on the humidity content, higher for saturated media

| Solid | E. coli | $H_2O$-liquid | $H_2O$-vapor | $O_2,CO_2$ | Drug |
|---|---|---|---|---|---|
| Structured (multileaf) vegetable | $0.01 - 1^*$ | | | | |
| Fresh potato under heat transfer | | $2 - 0.3^+$ | $2 - 0.3^+$ | | |
| Human tissues | | | | $5 \times 10^{-5}$ | $1 \times 10^{-5}$ |

## 5.1.6 Fick's Law Generalization

In complete analogy with the heat conduction in Sect. 2.1.3 (p. 17), a general form of Eqs. (5.5) or (5.17) can be written (with constant $D$) in terms of molar concentration $c$ of species 1 for vector mass flux $\mathbf{j}_1$ or $\hat{\mathbf{j}}_1$. For sake of simplicity, let us focus on the later case, by assuming that $c_1$ depend on more coordinates, as with a rectangular reference: $c_1 = c_1(x, y, z, \theta)$.

Let us take the surface that within the mixture connects all points having the same concentration: an **isoflux surface**. Then, let us consider two such surfaces, at a given time, having $c_1$ and $c_1 + dc_1$ concentrations, respectively, as well as the point $P$ on one of these (Fig. 5.3); the *species that flows by diffusion, in the unit of time, by unit area of isoflux surface* is the **mass flux** $\hat{\mathbf{j}}_{1n}$ **along the general n direction**. For an *isotropic medium*, $\hat{\mathbf{j}}_{1n}$ will be *normal* to the higher isoflux surface, and *oriented toward the lower isoflux surface*, while its *magnitude is proportional to the concentration gradient in the* **n** *direction*.

*In scalar terms*:

$$\hat{j}_{1n} = -D\frac{\partial c_1}{\partial \mathbf{n}}$$

(5.19)

while *in vector form*:

$$\hat{\mathbf{j}}_{1n} = -\nabla D c_1$$

(5.20)

**Fig. 5.3** Plane projection of
two isoconcentration curves,
which differs by an
infinitesimal value $dc_1$ and
of vector $\hat{\mathbf{j}}_{1n}$

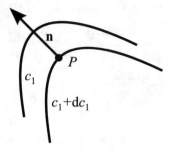

which is called the **generalized Fick's Law** *for an isotropic medium, not necessarily homogeneous.* We can also write in terms of mass concentration

$$\mathbf{j}_{1n} = -\nabla D\rho_1 \tag{5.21}$$

In rectangular Cartesian coordinates, the three components of $\hat{\mathbf{j}}_{1n}$ or $\mathbf{j}_{1n}$ for constant mass diffusivity are respectively as follows:

$$\hat{j}_{1x} = -D\frac{\partial c_1}{\partial x} \quad \hat{j}_{1y} = -D\frac{\partial c_1}{\partial y} \quad \hat{j}_{1z} = -D\frac{\partial c_1}{\partial z} \tag{5.22}$$

$$j_{1x} = -D\frac{\partial \rho_1}{\partial x} \quad j_{1y} = -D\frac{\partial \rho_1}{\partial y} \quad j_{1z} = -D\frac{\partial \rho_1}{\partial z} \tag{5.23}$$

## 5.2   Elementary Diffusion: Concentration Distribution in the Steady State, 1-D

In some cases in the practice, when the geometry and the mass driving forces are simple and the mixture is stationary, the concentration profiles and other related results can be directly computed based upon Fick's Law of diffusion. The simplest of such cases is the **species diffusion in the steady state through a 1-D shell (plane or cylindrical) geometry** $CV$**, without source term**, which will give the opportunity to define some important concept; knowing the concentration distribution is important to evaluate their response to the driving force for the mass transfer. We will see that the analytical development in these cases is analogous to some results of heat conduction.

In the first of such situations, as depicted in Fig. 5.4, we can compute the diffusion molar flux $\hat{j}_1$ in a stationary medium by using Eq. (5.17) similarly as already done with the conduction thermal flux $\dot{Q}_\lambda/A$ in Eq. (2.10)

$$\frac{\dot{Q}_\lambda}{A}\int_0^L dx = -\int_{T_1}^{T_2}\lambda_s dT \tag{2.10 revisited}$$

**Fig. 5.4** A substance layer which is infinite along the $x$- and $z$-directions, but with thickness $L$ along $y$-direction: indication of the concentration distribution $c_1(y)$, under a concentration difference $c_{1a} - c_{1b}$ between the limiting faces

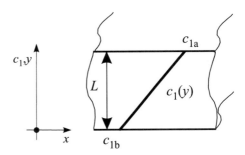

The integration limits are determined by observing that the left wall's face at $y = 0$ is at uniform concentration $c_{1a}$, with the right face at $y = L$ being at $c_{1b}$. Following the development of Sect. 2.2 (p. 20), one has:

$$\hat{j}_{1y} = -D\frac{dc_1}{dy} = D\frac{(c_{1a} - c_{1b})}{L} \tag{5.24}$$

or, equivalently, in terms of mass concentration, $\rho_1$, or mole fraction, $x_1$:

$$j_{1y} = -D\frac{d\rho_1}{dy} = D\frac{(\rho_{1a} - \rho_{1b})}{L} = D\frac{\rho M_1}{M}\frac{(x_{1a} - x_{1b})}{L} \tag{5.25}$$

Next, let us assume a *monolayer tube*, with *specified inner and outer concentrations*. A portion of the tube section is depicted in Fig. 5.5. The boundary conditions translate into the following:

$$c_1(r) = c_{1a}, \text{ for } r = r_a, \quad c_1(r) = c_{1b}, \text{ for } r = r_b \tag{5.26}$$

Performing the integration similarly than in Sect. 2.3.3.4 (p. 40), one has:

$$c_1(r) = c_{1a} - \frac{(c_{1a} - c_{1b})}{\ln\frac{r_b}{r_a}} \ln\frac{r}{r_a} \tag{5.27}$$

To calculate the diffusion molar flux $\hat{J}'_{1r}$ through the whole shell per unit axial length, we simply multiply the inner surface area when considering the diffusion molar flux $\hat{j}_{1r}$ applied to that surface, and recall Eq. (5.24) and Note 37 of Chap. 2:

$$\hat{J}'_{1r} \equiv \hat{j}_{1r}\bigg|_{r=r_a} 2\pi r_a = -D\frac{dc_1}{dr}\bigg|_{r=r_a} 2\pi r_a = \frac{D2\pi}{\ln\frac{r_b}{r_a}}(c_{1a} - c_{1b}) \tag{5.28}$$

while the corresponding mass flow rate per unit axial length $\dot{m}'_1$ is foud in the same way

$$\dot{m}'_1 = \frac{D2\pi}{\ln\frac{r_b}{r_a}}(\rho_{1a} - \rho_{1b}) \tag{5.29}$$

From what stated above, it will be $\dot{m}'_1 = M_1\hat{J}'_{1r}$.

**Fig. 5.5** *CV* portion of a
monolayer tube, with
indication of the limiting
concentrations

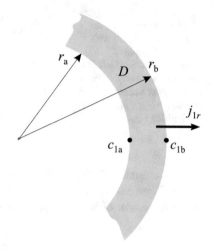

## 5.3   Fundamental Equation of Mass Transfer

So far we have dealt with the simplest cases of concentration distribution in the
steady state through a shell geometry. In analogy with heat conduction development,
Sect. 2.3 (p. 25), for more general geometries and in the unsteady regimes, we will
derive *generalized formulations*, that will find direct application for mass transport in
stationary as well as moving media (pure diffusion as well as mass convection). These
formulations can be applied, through the chemical or biological species balance, for
each *i* substance or species at stake.

Let us consider first the flow of mass through an infinitesimal *CV* fixed in a
Cartesian (rectangular) 2-D space, already represented in Fig. 2.10.

### 5.3.1   Equation of Species Conservation in Rectangular Coordinates

#### 5.3.1.1   The Balance on the Mass Concentration

Generally, species *i* is transported in a bulk substance 2 which moves itself, relative
to its reference frame. Recalling the definition of mass concentration $\rho_i$

$$\rho_i \equiv \frac{m_i}{V} \tag{5.1}$$

**Fig. 5.6** Infinitesimal $CV$, interested by species fluxes through its $CS$ (represented by black line arrows), and source/sink of mass species right within the same $CV$

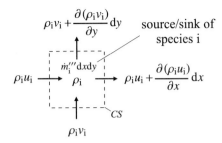

let us apply the component balance for a medium in motion on the $CV$ in terms of *fluxes* through its $CS$ and source/sink of component right within the same $CV$, while

$$\boxed{\rho_i \text{ entering } CS} + \boxed{\text{source/sink of } \rho_i \text{ in } CV, \text{ in the unit of time}}$$

$$= \boxed{\rho_i \text{ exiting } CS} + \boxed{\text{variation of } \rho_i \text{ in } CV, \text{ in the unit of time}} \qquad (5.30)$$

where the **source/sink term is the volumetric species flux** $\dot{m}_i'''$, in unit time and space (that is, the $CV$),[10] due to any PCB occurrence of interest.

Again, the volumetric source/sink term is taken as *uniform across the $CV$*; note that **it can be positive or negative, depending on the case at stake**: creation/consumption of chemical substances, as for the constituents of foodstuff during processing, and birth/death of evolving microorganisms, as for the colonization of bacteria or cell lines in structures of life science interest. For sake of generality, all these occurrences will be denoted as "species" in this development.

In Eq. (5.30), just as with Eq. (2.19), having written this source term on the left-hand side, when its sign is positive its contribution of species to the $CV$ is positive, as well. Following the development of Sect. 2.3.1 (p. 26), the terms of Eq. (5.30) can be written down as depicted in Fig. 5.6 and written out as follows:

$$-\left[\frac{\partial (\rho_i u_i)}{\partial x}dx + \frac{\partial (\rho_i v_i)}{\partial y}dy\right] \pm \dot{m}_i''' dxdy = (dxdy)\frac{\partial \rho_i}{\partial \theta}$$

Simplifying, we have

$$\frac{\partial \rho_i}{\partial \theta} + \frac{\partial (\rho_i u_i)}{\partial x} + \frac{\partial (\rho_i v_i)}{\partial y} = \pm \dot{m}_i''' \qquad (5.31)$$

---

[10] As with Note 20 of Chap. 2, the three primes $'''$ are purposeful to the fact that $\dot{m}_i'''$ refers to the *unit volume*. Therefore, $[\dot{m}_i'''] = \text{kg/m}^3$.

### 5.3.1.2  Species-, Mass-Averaged-, and Diffusion Velocities

Note that in the absence of chemical/biological reaction/transformation ($\dot{m}_i''' = 0$) at the right-hand side, Eq. (5.31) corresponds to the Equation of continuity

$$\frac{\partial \rho_f}{\partial \theta} + \frac{\partial (\rho_f u)}{\partial x} + \frac{\partial (\rho_f v)}{\partial y} = 0 \tag{3.54}$$

and for this purpose it is also called the *Equation of conservation of species i* (recall Note 21 in Chap. 3). Also note that, *regarding the occurrence as an ensemble* for all the *n* species (that is, taking the sum of all contributions), the sum of all the $\dot{m}_i'''$ terms is zero as **mass cannot be globally created or destroyed even in presence of reactions/transformations**:

$$\frac{\partial}{\partial \theta} \left( \sum_{i=1}^{n} \rho_i \right) + \frac{\partial}{\partial x} \left( \sum_{i=1}^{n} \rho_i u_i \right) + \frac{\partial}{\partial y} \left( \sum_{i=1}^{n} \rho_i v_i \right) = 0 \tag{5.32}$$

that is, we end up with the Equation of continuity again. Comparison of these two last Equations leads us to realize that, besides the definition of *density* of the mixture, the bulk or **mass-averaged velocity** components of the flow (*u* and *v*) arise as velocity component averages of all species velocity components in the mixture, indicating that the species can move relative to the mixture, even for a stationary one:

$$u = \frac{1}{\rho} \sum_{i=1}^{n} \rho_i u_i; \qquad v = \frac{1}{\rho} \sum_{i=1}^{n} \rho_i v_i \tag{5.33}$$

The velocity differences $u_i - u$ and $v_i - v$ are called the *diffusion velocities* of species *i* in the *x*- and *y*-direction, respectively. When the diffusion velocities are multiplied by the mass concentration $\rho_i$, we have the **mass flux** along the *k*-direction

$$\frac{\dot{m}_i}{A} \equiv j_{ik} \tag{5.3 revisited}$$

that is

$$j_{ix} \equiv \rho_i (u_i - u) \tag{5.34a}$$
$$j_{iy} \equiv \rho_i (v_i - v) \tag{5.34b}$$

Equation (5.31) then becomes:

$$\frac{\partial \rho_i}{\partial \theta} + \frac{\partial (\rho_i u)}{\partial x} + \frac{\partial (\rho_i v)}{\partial y} = -\frac{\partial j_{ix}}{\partial x} - \frac{\partial j_{iy}}{\partial y} \pm \dot{m}_i'''$$

For a constant density mixture flow, we can simplify the spatial derivatives at the left-hand side by using the Equation of continuity:

$$\frac{\partial \rho_i}{\partial \theta} + u\frac{\partial \rho_i}{\partial x} + v\frac{\partial \rho_i}{\partial y} = -\frac{\partial j_{ix}}{\partial x} - \frac{\partial j_{iy}}{\partial y} \pm \dot{m}_i''' \tag{5.35}$$

### 5.3.1.3 Rounding up

Now, let us plug down the Fick's Law for *uniform mass diffusivity*, or Eq. (5.23): we end up with the **fundamental Equation of mass transfer** with mass concentration $\rho_i$ as the independent variable, *in dimensional Cartesian form*:

$$\boxed{\frac{\partial \rho_i}{\partial \theta} + u\frac{\partial \rho_i}{\partial x} + v\frac{\partial \rho_i}{\partial y} = D_f\left(\frac{\partial^2 \rho_i}{\partial x^2} + \frac{\partial^2 \rho_i}{\partial y^2}\right) \pm \dot{m}_i'''} \tag{5.36}$$

Here, mass diffusivity $D_f$ refers to the species $i$, as it is convectively transported in the mixture. *In vector form*, we have

$$\boxed{\underbrace{\frac{D\rho_i}{D\theta}}_{①} = \underbrace{D_f \nabla^2 \rho_i}_{②} \underbrace{\pm \dot{m}_i'''}_{③}} \tag{5.37}$$

The various terms in Eq. (5.37) all represent mass fluxes (in the unit volume), *in competition* in every medium point: term ① represents the (transient) *convective mass flux*, term ② the *diffusion mass flux*, term ③ the *source/sink term* pertaining to reactive/transformational flows, with respect to chemical/biological species.

An equivalent form of fundamental Equation, with molar concentration $c_i$ as the independent variable, consistently with its definition

$$c_i = \frac{\rho_i}{M_i} \tag{5.16}$$

is written out as:

$$\boxed{\frac{Dc_i}{D\theta} = D_f \nabla^2 c_i \pm \frac{\dot{m}_i'''}{M_i}} \tag{5.38}$$

A fundamental Equation form in terms of molar fraction $x_i$ can be also found.

When trivially simplified, Eqs. (5.37) or (5.38) can be used for *stationary media*:

$$\boxed{\frac{\partial c_i}{\partial \theta} = D_s \nabla^2 c_i \pm \frac{\dot{m}_i'''}{M_i}} \tag{5.39}$$

*The system of the Navier–Stokes Equations*

$$\frac{\partial \rho}{\partial \theta} + (\nabla \cdot (\rho \mathbf{w})) = 0 \qquad\qquad \text{(3.70a revisited)}$$

$$\rho \left[ \frac{\partial \mathbf{w}}{\partial \theta} + (\mathbf{w} \cdot \nabla) \, \mathbf{w} \right] = \mu \nabla^2 \mathbf{w} - \nabla p \pm \mathbf{S} \qquad\qquad \text{(3.70b revisited)}$$

with the implied definition of mixture density $\rho$

$$\rho \equiv \sum_{i=1}^{n} \rho_i \qquad\qquad (5.9)$$

and supplemented by Eq. (5.37), together with the associate **initial conditions** and **boundary conditions**, *allows one to determine the functions* $\rho_i(x, y, \theta)$ or $c_i(x, y, \theta)$, $\mathbf{w}(x, y, \theta)$, and $p(x, y, \theta)$, in the specified assumptions, by means of numerical methods for any arbitrary geometry and state, as outlined in Sect. 5.5 (p. 236), that will be built upon the experience that we made so far by studying Sects. 2.5, 3.6, and 4.6 (pp. 55, 120, and 192, respectively).

### 5.3.2   BCs for the Fundamental Equation of Mass Transfer

The boundary conditions for mass transfer enjoy many analogies with conduction and convection heat transfer, but also deserve some special considerations. The relationships among the various variables reviewed in Sect. 5.1.4 (p. 207) apply. Generally, the following cases will occur:

- the subject medium consists of one phase, only. In this case, the mass transfer fundamental Eqs. (5.37, 5.38) can be coupled with some BCs notations earlier considered for conduction heat transfer, Sect. 2.3.1 (p. 26):

  1. when the concentration itself is specified, we have the first-kind BC;
  2. when the concentration gradient (mass flux) is specified, we have the second-kind BC. When nonzero, the *sign* of this quantity (the direction of mass flux with respect to the $CV$ at stake) *is always opposite of the sign of the concentration gradient.*

- a second neighboring phase can be recognized sharing an interface with the subject one. As we deduced in Sect. 4.5.4 (p. 190), one of the two approaches will be followed:

  *a segregated analysis:* modeling is restricted to the subject $CV$, regardless of the neighbouring medium sharing an interface with it. In this case, as important feature of mass transfer emerges: unlike temperature, *the species concentration does not vary continuously across the phase interface* [4]. Common examples are:

1.  the $CV$ at stake is a gas mixture bearing the species 1 as component is in thermodynamic contact (sharing pressure and temperature, for example) with the free surface of a species-1-pure liquid underneath, as water vapor in the air facing liquid water. The concentration distribution in the gas mixture will be determined by using, as first-kind BC at this interface, the partial (vapor) pressure $p_1$ that will be equal to its own saturation pressure at the temperature of the liquid $T_{\text{liq}}$ at the interface:

$$p_1 = p_{\text{sat}}(T_{\text{liq}}) \qquad (5.40)$$

2.  the $CV$ at stake is a liquid bearing a small amount of species 1 as solute (a dilute solution) is in thermodynamic contact with a gas mixture bearing the species 1 as component. The concentration distribution in the liquid will be determined by using, as first-kind BC at this interface, the mole fraction $x_1$ that will be equal to the partial pressure $p_1$ divided a constant, following *Henry's Law*:[11]

$$x_1 = \frac{p_1}{H} \qquad (5.41)$$

A collection of Henry's constant $H$ is reported in Table 5.5. Henry's Law applies at moderate to low pressures, only [4]: for example, when the partial pressure $p_1$ does not exceed 1 atm in the gas mixture. At higher pressures, $H$ is function of the partial pressure itself. An example of the use of Henry's Law is the depth- (pressure-) dependent dissolution of nitrogen in the blood of underwater divers. During careless decompression, the partial pressure of nitrogen becomes greater than the one in the atmosphere, leading to nitrogen evaporation (recalling, from Thermodynamics, an isothermal expansion in the phase diagram) and possible embolism.

*a conjugated analysis:* modeling is extended to the neighbouring, different phase. In this case, the concentration distribution can be determined simultaneously in both phases, with no need for assumptions at interface. In fact, the continuity of the concentration and mass flux are used as the conjugate boundary conditions to couple the mass equation for both phases, across their interface. In other words, with a conjugated mass convection analysis *the solution for concentration is shared seamlessly through the interface*. In this case, in complete analogy with convection heat transfer, we can speak of two-phase mass transfer modeling.

---

[11] As proposed by English chemist W. HENRY, at beginning XIX century.

**Table 5.5** Henry's constant $H$ (bar) for several gas at moderate pressures [4]

| $T$ (K) | Air | $N_2$ | $O_2$ | $CO_2$ | CO |
|---|---|---|---|---|---|
| 290 | $6.2 \times 10^4$ | $7.6 \times 10^4$ | $3.8 \times 10^4$ | $1.3 \times 10^3$ | $5.1 \times 10^4$ |
| 300 | $7.4 \times 10^4$ | $8.9 \times 10^4$ | $4.5 \times 10^4$ | $1.7 \times 10^3$ | $6.0 \times 10^4$ |
| 320 | $9.2 \times 10^4$ | $1.1 \times 10^5$ | $5.7 \times 10^4$ | $2.7 \times 10^3$ | $7.4 \times 10^4$ |
| 340 | $1.0 \times 10^5$ | $1.2 \times 10^5$ | $6.5 \times 10^4$ | $3.7 \times 10^3$ | $8.4 \times 10^4$ |

### 5.3.3  Other Coordinate Systems

In the assumptions of constant properties $D$ and $\rho$, for sake of simplicity with no source/sink term, the **fundamental Equation of mass diffusion in cylindrical coordinates** (Fig. 2.12) *in scalar notation* for the mass concentration $\rho_i$ is reported here:

$$\frac{\partial \rho_i}{\partial \theta} + v_r \frac{\partial \rho_i}{\partial r} + \frac{v_\phi}{r} \frac{\partial \rho_i}{\partial \phi} + v_z \frac{\partial \rho_i}{\partial z} = D_f \left[ \frac{1}{r} \frac{\partial}{\partial r} \left( r \frac{\partial \rho_i}{\partial r} \right) + \frac{1}{r^2} \frac{\partial^2 \rho_i}{\partial \phi^2} + \frac{\partial^2 \rho_i}{\partial z^2} \right] \qquad (5.42)$$

In the same assumptions, the **fundamental Equation of mass diffusion in spherical coordinates** can be written as follows:

$$\frac{D\rho_i}{D\theta} = D_f \left[ \frac{1}{r^2} \frac{\partial}{\partial r} \left( r^2 \frac{\partial \rho_i}{\partial r} \right) + \frac{1}{r^2 \sin \gamma} \frac{\partial}{\partial \gamma} \left( \sin \gamma \frac{\partial \rho_i}{\partial \gamma} \right) + \frac{1}{r^2 \sin^2 \gamma} \frac{\partial}{\partial \phi} \left( \frac{\partial \rho_i}{\partial \phi} \right) \right] \qquad (5.43)$$

### 5.3.4  Mass/Heat Transfer Correspondence for Stationary Media

In addition to steady transport through shells (Sect. 5.2, p. 212), the development of solution methods for the concentration distribution parallels what already seen for heat conduction, in particular with formal analytical methods, as in Sect. 2.3.3 (p. 31), for plates and cylinders in the steady state, with and without the source term $\dot{m}_i'''$. In the same way, the unsteady or transient mass diffusion, which is also relevant in many practical cases, such as the processes to a multitude of media in which species vary, parallels what already seen for heat conduction, as in Sect. 2.3.5 (p. 44).

In particular, the discussion leading to the *thermal* Biot number

$$\mathrm{Bi_T} \equiv \frac{\overline{h}_T L}{\lambda_s} \qquad (2.46 \text{ revisited})$$

brings to the definition of the **mass Biot number**

$$\boxed{Bi_M \equiv \frac{\overline{h}_M L}{D_s}} \tag{5.44}$$

informing on the competition between external mass convection and internal diffusion, as well as with the *thermal* Fourier number

$$Fo_T \equiv \frac{\theta \alpha_s}{L^2} \tag{2.94 revisited}$$

which can be transformed, to our scopes, to the **mass Fourier number**

$$\boxed{Fo_M \equiv \frac{\theta D_s}{L^2}} \tag{5.45}$$

seen as a dimensionless time, that is, the ratio between the physical time and a time constant of the process driven by diffusion.

As usual, the purpose of this analysis is to identify the controlling dimensionless parameters that are commonly used in these kind of applications. For all steady and unsteady cases, Table 5.6 summarizes the proper correspondence between heat and mass transfer variables, so that all results obtained with Sects. 2.3.3 (p. 31) to 2.3.5 (p. 44) can be extended to mass transfer for stationary media. With these positions, it is straightforward to transform all analytical or approximate results of heat transfer into the corresponding results of mass transfer. As examples taken from the conduction solutions in the unsteady state:

In Initial regime—semi-infinite plate model, first-kind BC We can use

$$\frac{T(x, \theta) - T_\infty}{T_i - T_\infty} = erf\left[\frac{x}{2\sqrt{\alpha_s \theta}}\right] \tag{2.70 revisited}$$

to yield (with Note 39 of Chap. 2 in mind)

$$\frac{c_1(x, \theta) - c_{1\infty}}{c_{1i} - c_{1\infty}} = erf\left[\frac{x}{2\sqrt{D_s \theta}}\right] \tag{5.46}$$

In Late regime—lumped model We can use

$$\frac{T(\theta) - T_\infty}{T_i - T_\infty} = exp\left(-\frac{\overline{h}A}{\rho_s c_{ps} V}\theta\right) \tag{2.78 revisited}$$

**Table 5.6** Correspondence between heat and mass transfer variables

| Temperature | | Concentration |
|---|---|---|
| $T$ | $\rightarrow$ | $c_1, \rho_1$ |
| $T_0$ | $\rightarrow$ | $c_{10}, \rho_{10}$ |
| $T_p$ | $\rightarrow$ | $c_{1p}, \rho_{1p}$ |
| $T_i$ | $\rightarrow$ | $c_{1i}, \rho_{1i}$ |
| $T_\infty$ | $\rightarrow$ | $c_{1\infty}, \rho_{1\infty}$ |
| $\alpha_s$ | $\rightarrow$ | $D_s$ |
| $\lambda_s$ | $\rightarrow$ | $D_s$ |
| $\rho_s$ | $\rightarrow$ | not available |
| $c_{ps}$ | $\rightarrow$ | not available |
| $h_T$ | $\rightarrow$ | $h_M$ |
| $Bi_T$ | $\rightarrow$ | $Bi_M$ |
| $Fo_T$ | $\rightarrow$ | $Fo_M$ |
| $\dot{q}_h$ | $\rightarrow$ | $\hat{j}_{1p}, j_{1p}$ |
| $\dot{e}'''/\lambda_s$ | $\rightarrow$ | $\dot{m}_i'''/M_i, \dot{m}_i'''$ |

to yield

$$\frac{c_1(\theta) - c_{1\infty}}{c_{1i} - c_{1\infty}} = \exp\left(-\frac{\overline{h}_M A}{V}\theta\right) \qquad (5.47)$$

In General unsteady regime—dimensionless solution  We can use

$$T^* = T^*\left(x^*, Fo_T, Bi_T\right) \qquad (2.102 \text{ revisited})$$

to yield

$$c_1^* = c_1^*\left(x^*, Fo_M, Bi_M\right) \qquad (5.48)$$

## 5.3.5  Turbulent Mass Transfer

Like turbulent flow and heat transfer, turbulent mass transfer is encountered in many practical applications and all notations and relationships follow the development of Sect. 4.3.4 (p. 167): the ideally measured concentration of a substance in one point in a given $CV$ of a turbulent flow will undergo some irregular, rapid fluctuation about a mean value, with the actual concentration value for species 1, $\overline{c}_1$, regarded as the sum of the mean (time-averaged) value and its fluctuation. We have therefore the time-smoothed version of the Equation of mass transfer by diffusion Eq. (5.38):

$$\left( \frac{\partial \overline{c_1}}{\partial \theta} + \frac{\partial \overline{c_1}\,\overline{u}}{\partial x} + \frac{\partial \overline{c_1}\,\overline{v}}{\partial y} + \frac{\partial \overline{c_1}\,\overline{w}}{\partial z} \right) +$$

$$\underbrace{\left( \frac{\partial \overline{c_1' u'}}{\partial x} + \frac{\partial \overline{c_1' v'}}{\partial y} + \frac{\partial \overline{c_1' w'}}{\partial z} \right)}_{\text{turb. mass transport due to fluctuations}} = D_f \nabla^2 \overline{c_1} \pm \frac{\overline{\dot{m}_1'''}}{M_1} \qquad (5.49)$$

There the under-braced terms appear, which describe **the mass transport associated with the turbulent fluctuations, which adds to the molecular transport**.

The eddy diffusivity concept is exploited once more with the *mass eddy diffusivity* $\varepsilon_D$ (with $[\varepsilon_D]=$m$^2$/s), so to evidence the added contribution of turbulence to molecular heat transport. Letting

$$- \overline{c_1' w_j'} \equiv \varepsilon_D \frac{\partial \overline{c_1}}{\partial x_j} \qquad (5.50)$$

Equation (5.49) becomes

$$\boxed{\frac{\partial \overline{c_1}}{\partial \theta} + \frac{\partial}{\partial x_j}\left( \overline{c_1 w_j} \right) = \frac{\partial}{\partial x_j}\left( (D_f + \varepsilon_D) \frac{\partial \overline{c_1}}{\partial x_j} \right) \pm \frac{\overline{\dot{m}_1'''}}{M_1}} \qquad (5.51)$$

Unlike molecular thermal diffusivity, which is a parameter of the fluid, the mass eddy diffusivity $\varepsilon_D$ depends strongly on the position within the flow, while going to zero at the flow-limiting wall. Its value is approximately the same as the momentum eddy diffusivity $\varepsilon_v$ we dealt with in Sect. 3.3.7.4 (p. 104).

Finally, it should be considered that in many important application cases such as combustion, the *turbulence intertwines with chemistry*, and supplementary contributions to the volumetric species source/sink term $\overline{\dot{m}'''}$ due to turbulence may occur.[12] However, these cases are the subject of a vast field of specific studies.

---

[12] The simplest cases for last term in Eq. (5.51) are a *first-* or *second-order* reaction/transformation [5]. In these cases, considering a uniform and constant rate $K$ (1/s) for sake of simplicity, after time-averaging one gets:

$$\frac{\overline{\dot{m}_1'''}}{M_1} = \begin{cases} K_I \overline{c_1}, & \text{first-order reaction} \\ K_{II}(\overline{c_1}^2 + \underbrace{\overline{c_1'^2}}), & \text{second-order reaction} \end{cases}$$

where the bracketed term appears due to turbulence, marking the difference between reactions of first order and second order.

## 5.4    Dimensionless Equation and Interactions in Convection Mass Transfer

We put in proper evidence so far that mass transfer shares many modeling aspects with heat transfer, although with some due differences. Therefore, the discussion on the identification of the controlling dimensionless parameters that are commonly used in the applications of mass transfer, also providing an introduction to and correlations for this mechanism, parallels that of Sect. 4.4 (p. 168).

### 5.4.1   Dimensional Analysis of Convection Mass Transfer

To complete the dimensional analysis of transport phenomena, let us start by invoking the four governing Equations for species 1 in a *binary, laminar, incompressible fluid* with *constant rheological properties*. Now, the opportunity is seized to develop a **most general form of solution**, which will include *every possible source/sink term* for each transport mechanism and convection regime. The final goal, as usual, is to solve for velocity $\mathbf{w}$, pressure $p$, temperature $T$, and mass concentration $\rho_1$ (for the general $CV$ of Fig. 4.15, p. 173):

1. we will start by recalling the *continuity Equation* as follow:

$$(\nabla \cdot \mathbf{w}) = 0 \qquad \text{(3.70a revisited)}$$

2. next, we realize that the momentum conservation would need to include *two separate buoyancy force terms*: one resulting from a temperature gradient, the other for a concentration gradient in the fluid. We recall that, to take into account the Equation of state $\rho_f = \rho_f(p, T)$ we used Taylor's series to yield

$$\rho_f(T) = \rho_{f\infty} + \left(\frac{\partial \rho_f}{\partial T}\right)_p (T - T_\infty) + \left(\frac{\partial \rho_f}{\partial p}\right)_T (p - p_\infty) + \text{higher-order terms} \tag{4.64}$$

Now, with the Equation of state being $\rho_f = \rho_f(p, T, \rho_1)$, we can extend the reasoning leading to

$$\frac{\rho_{f\infty} - \rho_f}{\rho_f} = -\frac{1}{\rho_f}\left(\frac{\partial \rho_f}{\partial T}\right)_p (T - T_\infty) = \beta(T - T_\infty) \tag{4.65}$$

so that

$$\frac{\rho_{f\infty} - \rho_f}{\rho_f} = -\frac{1}{\rho_f}\left(\frac{\partial \rho_f}{\partial T}\right)_p (T - T_\infty) - \frac{1}{\rho_f}\left(\frac{\partial \rho_f}{\partial \rho_1}\right)_p (\rho_1 - \rho_{1\infty})$$

$$= \beta_T(T - T_\infty) + \beta_M(\rho_1 - \rho_{1\infty}) \tag{5.52}$$

using again the coefficient of thermal expansion

$$\beta \equiv \beta_T \equiv -\frac{1}{\rho_f}\left(\frac{\partial \rho_f}{\partial T}\right)_p \tag{4.40 revisited}$$

and a new coefficient relating density to the variation of composition

$$\beta_M \equiv -\frac{1}{\rho_f}\left(\frac{\partial \rho_f}{\partial \rho_1}\right)_p \tag{5.53}$$

So we can write the *conservation of momentum Equation* as follows:

$$\frac{D\mathbf{w}}{D\theta} = \nu\nabla^2\mathbf{w} - \frac{1}{\rho_f}\nabla p + \mathbf{g}\beta_T(T - T_\infty) + \mathbf{g}\beta_M(\rho_1 - \rho_{1\infty}) \tag{5.54}$$

3. then we can recall the *conservation of energy Equation* as follows:

$$\frac{1}{\alpha_f}\frac{DT}{D\theta} = \nabla^2 T \pm \frac{\dot{e}'''}{\lambda_f} \tag{4.38 revisited}$$

4. and finally and *conservation of mass for the species 1 Equation* as follows:

$$\frac{D\rho_1}{D\theta} = D_f\nabla^2\rho_1 \pm \dot{m}_1''' \tag{5.37}$$

Then, let us choose some suitable dimensionless independent and dependent, relative to a flow having reference temperature, velocity, pressure, and mass concentration of $T_\infty$, $\mathbf{w}_\infty$, $\rho_f\mathbf{w}_\infty^2$, and $\rho_{1\infty}$, respectively. The flow features an arbitrary reference temperature difference $\Delta T$ and an arbitrary reference mass concentration difference $\Delta\rho_1$, and interacts with a generic geometry having a reference length $L$. We can choose therefore:

$$\boxed{\begin{array}{l} x_j^* \equiv \dfrac{x_j}{L}, \quad \nabla^* \equiv L\nabla, \quad \dfrac{D}{D\theta^*} \equiv \dfrac{L}{\mathbf{w}_\infty}\dfrac{D}{D\theta}, \quad p^* \equiv \dfrac{p}{\rho_f\mathbf{w}_\infty^2}, \quad \mathbf{w}^* \equiv \dfrac{\mathbf{w}}{\mathbf{w}_\infty}, \\[3mm] \qquad\qquad T^* \equiv \dfrac{T - T_\infty}{\Delta T}, \quad \rho_1^* \equiv \dfrac{\rho_1 - \rho_{1\infty}}{\Delta\rho_1} \end{array}} \tag{5.55}$$

We duly recall that $\mathbf{w}$ is the bulk or mass-averaged velocity of the flow, having components such as

$$u = \frac{1}{\rho} \sum_{i=1}^{n} \rho_i u_i; \qquad v = \frac{1}{\rho} \sum_{i=1}^{n} \rho_i v_i \qquad (5.33)$$

We start by operating the change of variable as performed already in Sect. 3.4 (p. 108),

1. starting with for the **dimensionless Equation of continuity**, to obtain again

$$\boxed{\left(\nabla^* \cdot \mathbf{w}^*\right) = 0} \qquad \text{(3.104a revisited)}$$

2. then we proceed with the **dimensionless momentum conservation Equation**, to obtain

$$\frac{D\mathbf{w}^*}{D\theta^*} = \frac{1}{\mathrm{Re}_L} \nabla^{*2}\mathbf{w}^* - \nabla^* p^* + \frac{\mathrm{Gr_T}}{\mathrm{Re}_L^2} \frac{\mathbf{g}}{g} T^* + \frac{\mathrm{Gr_M}}{\mathrm{Re}_L^2} \frac{\mathbf{g}}{g} \rho_1^*$$

having divided both hand sides by $\mathbf{w}_\infty^2 / L$, or

$$\boxed{\frac{D\mathbf{w}^*}{D\theta^*} = \frac{1}{\mathrm{Re}_L} \nabla^{*2}\mathbf{w}^* - \nabla^* p^* + \frac{\mathrm{Ri_T}\mathbf{g}}{g} T^* + \frac{\mathrm{Ri_M}\mathbf{g}}{g} \rho_1^*} \qquad (5.56)$$

3. we continue with the **dimensionless energy conservation Equation**

$$\boxed{\frac{DT^*}{D\theta^*} = \frac{1}{\mathrm{Re}_L \mathrm{Pr}} \nabla^{*2} T^* \pm \frac{\mathrm{Po}_L}{\mathrm{Re}_L \mathrm{Pr}}} \qquad (5.57)$$

having divided both hand sides by $\mathbf{w}_\infty \Delta T / (\alpha_{\mathrm{f}} L)$, and simplified last term multiplying and dividing by $\nu$

4. finally, we complete with the **dimensionless mass conservation Equation for species 1**

$$\boxed{\frac{D\rho_1^*}{D\theta^*} = \frac{1}{\mathrm{Re}_L \mathrm{Sc}} \nabla^{*2} \rho_1^* \pm \mathrm{Da}_L} \qquad (5.58)$$

having divided both hand sides by $\mathbf{w}_\infty \Delta \rho_1 / L$.

Here, beside the Reynolds number

$$\mathrm{Re}_L \equiv \frac{\rho_{\mathrm{f}} \mathbf{w}_\infty L}{\mu} \qquad \text{(3.2 revisited)}$$

and the Prandtl number

$$\mathrm{Pr} \equiv \frac{\nu}{\alpha_\mathrm{f}} \qquad \text{(4.50 revisited)}$$

the following new *dimensionless numbers* have been defined:[13]

- **thermal Richardson number** expresses the *ratio of thermal buoyancy effects to forced flow effects*:[14]

$$\boxed{\mathrm{Ri}_{LT} \equiv \frac{\mathrm{Gr}_{LT}}{\mathrm{Re}_L^2}} \qquad (5.59)$$

involving the *thermal* Grashof number

$$\mathrm{Gr}_{LT} \equiv \frac{g\beta_\mathrm{T}\Delta T L^3}{\nu^2} \qquad \text{(4.70 revisited)}$$

- **mass Richardson number** expresses the *ratio of concentration buoyancy effects to forced flow effects*:

$$\boxed{\mathrm{Ri}_{LM} \equiv \frac{\mathrm{Gr}_{LM}}{\mathrm{Re}_L^2}} \qquad (5.60)$$

involving the **mass Grashof number**, analogous to the $\mathrm{Gr}_{LT}$, arising due to the *buoyant force caused by concentration gradients*:

$$\boxed{\mathrm{Gr}_{LM} \equiv \frac{g\beta_\mathrm{M}\Delta T L^3}{\nu^2}} \qquad (5.61)$$

---

[13]The $\mathrm{Ri_T}$, Po, Sc, and Da numbers were, respectively, proposed by the English mathematician, physicist, and meteorologist L. FRY RICHARDSON, by the Russian engineer A.A. POMERANT'SEV, by the German engineer E. SCHMIDT, and by German physical chemist G.F. DAMKÖHLER, all in mid-twentieth century.

[14]So far, we held the thermal Grashof number $\mathrm{Gr_T}$ as an appropriate dimensionless group in natural convection. With Note 20 of Chap. 4 in mind, we know that this assumption may be criticized. Without digging in how this would reflect in the solution of natural convection configurations, we acknowledge now that the thermal Richardson number $\mathrm{Ri_T}$ comes in handy as a criterion for transition from one convection regime to the other. As summarized by Sarghini, F., Ruocco, G.: Enhancement and reversal heat transfer by competing modes in jet impingement. International Journal of Heat and Mass Transfer (2004) doi: 10.1016/j.ijheatmasstransfer.2003.10.015, one has, depending on Pr:

$$\mathrm{Ri_T^{0.25}} \begin{cases} > 1, \text{ natural convection} \\ < 1, \text{ forced convection} \end{cases} \text{ for fluids with } \mathrm{Pr} < 1$$

and

$$\frac{\mathrm{Ri_T^{0.25}}}{\mathrm{Pr^{0.083}}} \begin{cases} > 1, \text{ natural convection} \\ < 1, \text{ forced convection} \end{cases} \text{ for fluids with } \mathrm{Pr} > 1$$

- **Pomerantsev number** characterizes the *heat source due to a PCB phenomenon compared to the medium thermal conductivity*:

$$\mathrm{Po}_L \equiv \frac{\dot{e}''' L^2}{\lambda_f \Delta T} \tag{5.62}$$

involving the volumetric energy flux $\dot{e}'''$ we dealt with in Fig. 2.11 (p. 28).
- **Schmidt number** represents the *relative importance of molecular momentum transfer over molecular mass transfer*:

$$\mathrm{Sc} \equiv \frac{\nu}{D_f} \tag{5.63}$$

- **Damkohler number** compares the *mass source due to a PCB phenomenon to the mass transported by the flow*, through the ratio of the related characteristic times $L/\mathbf{w}_\infty$ and $\Delta\rho_1/\dot{m}_1'''$:

$$\mathrm{Da}_L \equiv \frac{\dot{m}_1''' L}{\mathbf{w}_\infty \Delta\rho_1} = \frac{\Delta\theta_{\mathrm{transport}}}{\Delta\theta_{\mathrm{PCB}}} \tag{5.64}$$

involving the volumetric species flux $\dot{m}'''$ we dealt with in Fig. 5.6 (p. 238).

In the end, we see that the intertwined functions **in dimensional form** we sought were

$$\mathbf{w} = \mathbf{w}\left(x_j, L, \rho_f, \nu, T, T_\infty, \rho_{1\infty}, \mathbf{w}_\infty\right) \text{ and } p = p\left(x_j, L, \rho_f, \nu, T, T_\infty, \rho_{1\infty}, \mathbf{w}_\infty\right) \tag{5.65}$$

$$T = T\left(x_j, L, \alpha_f, \lambda_f, T_\infty, \Delta T, \dot{e}''', \mathbf{w}, \mathbf{w}_\infty\right) \tag{5.66}$$

$$\rho_1 = \rho_1\left(x_j, L, D_f, \rho_{1\infty}, \Delta\rho_1, \dot{m}_1''', \mathbf{w}, \mathbf{w}_\infty\right) \tag{5.67}$$

while we have obtained, beside

$$\mathbf{w}^* = \mathbf{w}^*\left(x_j^*, T^*, \mathrm{Re}_L, \mathrm{Ri}_{LT}, \mathrm{Ri}_{LM}\right) \text{ and } p^* = p^*\left(x_j^*, T^*, \mathrm{Re}_L, \mathrm{Ri}_{LT}, \mathrm{Ri}_{LM}\right) \tag{5.68}$$

$$T^* = T^*\left(x_j^*, \mathbf{w}^*, \mathrm{Re}_L, \mathrm{Pr}, \mathrm{Po}_L\right) \tag{5.69}$$

also the following solution, also **in dimensionless form**:

$$\rho_1^* = \rho_1^*\left(x_j^*, \mathbf{w}^*, \mathrm{Re}_L, \mathrm{Sc}, \mathrm{Da}_L\right) \tag{5.70}$$

with the resulting ease of discussion.

## 5.4.2 Boundary Layers and the Heat/Mass Transfer Similitude

With mass and heat transfer yielding for easily recognized correspondences, we found similitudes even when the boundary layer concept is recalled, for example, for the simplest one illustrated by Fig. 5.7 (in parallel with the ones shown in Figs. 4.16, 4.17). Therefore, *temperature and concentration boundary layers are governed by the same phenomena* [4]: a concentration boundary layer would be described by a development of the boundary layer thickness $\delta_M$, similarly than the reasoning made for $\delta_T$. In particular, the introduction of the Schmidt number Eq. (5.63) provides the opportunity to review

$$\frac{\delta}{\delta_T} \simeq Pr^n \qquad (4.98)$$

in terms of comparison between velocity and concentration boundary layers:

$$\frac{\delta}{\delta_M} \simeq Sc^n \qquad (5.71)$$

where for most applications it is again reasonable to assume a value of $n = 1/3$ [6].

As noted earlier, this similitude can be profitably exploited by simply adopting the heat transfer solution which is thought to be applicable in the given configuration and operation. Along the same lines that brought us to Table 5.6, we can establish that there is a precise similitude between other driving parameters between the two mechanisms, as reviewed in Table 5.7 and, by changing the notation accordingly, we can deduce the proper solution for convection mass transfer.

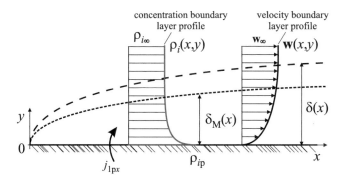

**Fig. 5.7** Flow parallel to a plate: the concentration and velocity distributions, $\rho_i(x, y)$ and $\mathbf{w}(x, y)$, being deformed along $x$, in the developing the boundary layer. The boundary layers are indicated by the $x$−varying thicknesses $\delta_M$ and $\delta$. The mass flux $j_{1px}$ exchanged locally between plate and flow is also indicated

| **Table 5.7** Comparison between temperature and concentration boundary layers | Temperature boundary layer | | Concentration boundary layer |
|---|---|---|---|
| | $T$ | $\rightarrow$ | $c_1, \rho_1$ |
| | $T_\mathrm{p}$ | $\rightarrow$ | $c_{1\mathrm{p}}, \rho_{1\mathrm{p}}$ |
| | $T_\infty$ | $\rightarrow$ | $c_{1\infty}, \rho_{1\infty}$ |
| | $\alpha_\mathrm{f}$ | $\rightarrow$ | $D_\mathrm{f}$ |
| | $\lambda_\mathrm{f}$ | $\rightarrow$ | $D_\mathrm{f}$ |
| | $\dot{q}_h$ | $\rightarrow$ | $\hat{\jmath}_{1\mathrm{p}}, j_{1\mathrm{p}}$ |
| | $\mathrm{Nu}_x$ | $\rightarrow$ | $\mathrm{Sh}_x$ |
| | $\mathrm{Pr}$ | $\rightarrow$ | $\mathrm{Sc}$ |

As an example, let us consider the case of *external forced **heat** convection along a flat plate* which, before full boundary layer development and for *high Prandtl numbers and uniform surface temperature* can be solved analytically yielding the following local heat transfer notation:

$$\mathrm{Nu}_x = 0.332\,\mathrm{Pr}^{1/3}\mathrm{Re}_x^{1/2} \quad \text{for} \ \ \mathrm{Pr} \gtrsim 0.5 \qquad \text{(4.99a revisited)}$$

that, with the definition of local Nusselt number

$$|\mathrm{Nu}_x| \equiv \frac{\dot{q}_{hx}x}{\Delta T \lambda_\mathrm{f}} \qquad (4.85)$$

may be revisited again as

$$\frac{\dot{q}_{hx}x}{\Delta T \lambda_\mathrm{f}} = 0.332 \left(\frac{\nu}{\alpha_\mathrm{f}}\right)^{1/3}\left(\frac{\mathbf{w}_\infty x}{\nu}\right)^{1/2} \quad \text{for} \ \ \left(\frac{\nu}{\alpha}\right) \gtrsim 0.5 \qquad \text{(4.99a revisited)}$$

Then, with the notation change of Table 5.7, we can get the solution for the *external forced **mass** convection along a flat plate* before full boundary layer development and for *high Schmidt numbers and uniform surface concentration*

$$\frac{\hat{\jmath}_{1\mathrm{p}}x}{\Delta c\,D_\mathrm{f}} = 0.332 \left(\frac{\nu}{D_\mathrm{f}}\right)^{1/3}\left(\frac{\mathbf{w}_\infty x}{\nu}\right)^{1/2} \quad \text{for} \ \ \left(\frac{\nu}{D_\mathrm{f}}\right) \gtrsim 0.5 \qquad (5.72)$$

or, by defining the **local Sherwood number**[15]

$$\boxed{\ \mathrm{Sh}_x \equiv \frac{\hat{\jmath}_{1\mathrm{p}}x}{\Delta c\,D_\mathrm{f}} \equiv \frac{h_\mathrm{M}L}{D_\mathrm{f}}\ } \qquad (5.73)$$

---

[15] As proposed by the American chemical engineer T.K. SHERWOOD in mid-twentieth century.

we have

$$\mathrm{Sh}_x = 0.332 \mathrm{Sc}^{1/3} \mathrm{Re}_x^{1/2} \quad \text{for} \quad \mathrm{Sc} \gtrsim 0.5 \tag{5.74}$$

In the same way, we can derive for the other range of Sc number

$$\mathrm{Sh}_x = 0.564 \mathrm{Sc}^{1/2} \mathrm{Re}_x^{1/2} \quad \text{for} \quad \mathrm{Sc} \lesssim 0.5 \tag{5.75}$$

being $\mathrm{Pe} = \mathrm{PrRe}$.

As we defined the average Nusselt numbers by

$$\overline{\mathrm{Nu}}_L \equiv \frac{\overline{h}_T L}{\lambda_{\mathrm{f}}} \tag{4.56a}$$

then we can postulate that

$$\overline{\mathrm{Sh}}_L \equiv \frac{\overline{h}_M L}{D_{\mathrm{f}}} \tag{5.76}$$

and hence

$$\overline{\mathrm{Sh}}_L = 1.13 \mathrm{Sc}^{1/2} \mathrm{Re}_L^{1/2} \quad \text{for} \quad \mathrm{Pr} \lesssim 0.5 \tag{5.77}$$

$$\overline{\mathrm{Sh}}_L = 0.664 \mathrm{Sc}^{1/2} \mathrm{Re}_L^{1/2} \quad \text{for} \quad \mathrm{Pr} \gtrsim 0.5 \tag{5.78}$$

### 5.4.3 Analogies Among Transport Phenomena

We had some opportunity so far to infer on the strong similarities between molecular heat, mass, and momentum transfer. Similarities in turbulent transport also exist, as we were able to develop the governing Eqs. (3.91, 4.47, 5.51) (with no source/sink terms, for sake of simplicity)

$$\frac{\partial \overline{w}_i}{\partial \theta} + \frac{\partial}{\partial x_j}\left(\overline{w}_i \overline{w}_j\right) = \frac{\partial}{\partial x_j}\left((\nu + \varepsilon_\nu)\frac{\partial \overline{w}_i}{\partial x_j}\right) - \frac{1}{\rho_{\mathrm{f}}}\frac{\partial \overline{p}}{\partial x_i} \tag{3.90 revisited}$$

$$\frac{\partial \overline{T}}{\partial \theta} + \frac{\partial}{\partial x_j}\left(\overline{T}\overline{w}_j\right) = \frac{\partial}{\partial x_j}\left((\alpha_{\mathrm{f}} + \varepsilon_\alpha)\frac{\partial \overline{T}}{\partial x_j}\right) \tag{4.47}$$

$$\frac{\partial \overline{c_1}}{\partial \theta} + \frac{\partial}{\partial x_j}\left(\overline{c_1 w_j}\right) = \frac{\partial}{\partial x_j}\left((D_{\mathrm{f}} + \varepsilon_D)\frac{\partial \overline{c_1}}{\partial x_j}\right) \tag{5.51 revisited}$$

showing an increment of the molecular transport for each mechanism at stake. Even at first glance the similarities are evident, but other successful approaches have been worked out on this topic to develop feasible formulas, so as to allow prediction of one transport mechanism from another one.

First, we realize that a **relationship between turbulent wall shear stress and convection heat flux** exists. Assuming the simplest of configurations, that is, the flow parallel to the plate, we recall Figs. 4.16 and 4.17 in Sect. 4.5.1 (p. 177) with which we inferred that the two velocity and temperature thickness distributions exist. This relationship holds as well in case of turbulent flows. In particular, we saw that $\delta_T = \delta$ implied

$$Pr \simeq 1 \qquad (4.96)$$

Now, the structure of the mean (time-averaged) temperature and concentration distributions in a turbulent boundary layer mimics that of the velocity longitudinal component. For this reason, "turbulent" versions of Eqs. (3.5, 2.7)

$$\tau_{yx,x} = -\mu \left. \frac{du}{dy} \right|_{y=0} \qquad \text{(3.5 revisited)}$$

$$\dot{q}_\lambda = -\lambda \left. \frac{dT}{dy} \right|_{y=0} \qquad \text{(2.7 revisited)}$$

are considered valid, when written with reference to some "turbulent" shear stress and convective heat flux right at the wall:

$$\tau_t = -\rho_f (\nu + \varepsilon_\nu) \frac{d\overline{u}}{dy} \qquad (5.79)$$

$$\dot{q}_t = -\rho_f c_{p_f} (\alpha_f + \varepsilon_\alpha) \frac{d\overline{T}}{dy} \qquad (5.80)$$

Dividing Eqs. (5.79, 5.80) side by side

$$\frac{\tau_t}{\dot{q}_t} = \frac{\rho_f (\nu + \varepsilon_\nu)}{\rho_f c_{p_f} (\alpha_f + \varepsilon_\alpha)} \frac{d\overline{u}}{d\overline{T}}$$

and assuming that $\nu = \alpha_f$ and $\varepsilon_\nu = \varepsilon_\alpha$ (i.e., Pr = 1 and $Pr_t = 1$), we get

$$\frac{\tau_t}{\dot{q}_t} c_{p_f} d\overline{T} = d\overline{u} \qquad (5.81)$$

With this analogy, this result will hold for any transversal position in the turbulent boundary layer and the ratio at the left-hand side will be constant for any $y$. Then, we integrate between conditions at the plate wall where $\overline{T} = T_p$ and $\overline{u} = 0$ to a large enough $y$ where $\overline{T} \simeq T_\infty$ and $\overline{u} \simeq |\mathbf{w}_\infty|$, that is the reference velocity, having

$$\frac{\tau_t}{\dot{q}_t} = \frac{|\mathbf{w}_\infty|}{c_{p_f}\left(T_\infty - T_p\right)} \tag{5.82}$$

Now, we can exploit two definitions that still hold with this turbulent boundary layer: the *local skin friction coefficient* $C_{fx}$

$$C_{fx} \equiv \frac{\tau_t}{\frac{1}{2}\rho_f|\mathbf{w}_\infty|^2} \tag{3.122 revisited}$$

and the *local heat transfer coefficient* $h_T$

$$h_T \equiv \frac{\dot{q}_t}{\left(T_\infty - T_p\right)} \tag{4.86 revisited}$$

so that we define the following new *dimensionless number*[16]

- **local thermal Stanton number**, expressing the *ratio of the heat transferred to a fluid to its thermal capacity* (valid also in laminar regime)

$$\boxed{St_{Tx} \equiv \frac{h_T}{\rho_f c_{pf}|\mathbf{w}_\infty|} \equiv \frac{\dot{q}_x}{\rho_f c_{pf}|\mathbf{w}_\infty|\left(T_\infty - T_p\right)} \equiv \frac{Nu_x}{Pe_x} \equiv \frac{Nu_x}{Re_x Pr}} \tag{5.83}$$

This definition implies another one

- **local mass Stanton number**, expressing the *ratio of the mass transfer form/to a solid phase surface to the mass transferred by a fluid phase*

$$\boxed{St_{Mx} \equiv \frac{h_M}{|\mathbf{w}_\infty|} \equiv \frac{Sh_x}{Re_x Sc}} \tag{5.84}$$

With these two definitions we can come up with[17]

$$\boxed{St_{Tx} = \frac{1}{2}C_{fx} \quad \text{for} \quad Pr \simeq 1, \qquad St_{Mx} = \frac{1}{2}C_{fx} \quad \text{for} \quad Sc \simeq 1} \tag{5.85}$$

The preceding analysis holds strictly for $Pr = 1$ (and $Pr_t = 1$) and $Sc = 1$. For different Pr numbers, *the empirical Colburn analogy* is useful[18]:

$$\boxed{St_{Tx}Pr^{2/3} = \frac{1}{2}C_{fx} \quad \text{for} \quad Pr \gtrsim 0.6} \tag{5.86}$$

---

[16] As proposed by English engineer T.E. STANTON at beginning twentieth century.

[17] Also known as *the Reynolds analogy*, in recognition of the work by the aforementioned scientist O. REYNOLDS.

[18] As proposed by American chemical engineer A.P. COLBURN at mid-twentieth century.

As a consequence, we can come up with a most widely used **generalized relationship between heat, momentum, and mass transfer**, valid **for both laminar and turbulent regimes**[19] called the *Chilton-Colburn J-factor analogy*. to correlate data from the three different transport mechanisms. It is written as follows:

$$J_v = \frac{C_f}{2} = J_h = St_T Pr^{2/3} = J_D = St_M Sc^{2/3} \tag{5.87}$$

Finally, the notations introduced with

$$\frac{\delta}{\delta_T} \simeq Pr^n \tag{4.98}$$

and

$$\frac{\delta}{\delta_M} \simeq Sc^n \tag{5.71}$$

allow one to define another dimensionless fluid property, which is related to Pr and Sc:

- **Lewis number**, that is relevant to evaluate boundary thicknesses in any situation involving simultaneous heat and mass transfer by convection:

$$Le \equiv \frac{\alpha}{D_f} = \frac{Sc}{Pr} \tag{5.88}$$

from which it follows that

$$\frac{\delta_T}{\delta_M} \simeq Le^n \tag{5.89}$$

where it is again reasonable to assume $n = 1/3$ [6].

As the dimensional analysis has been of concern in many parts of this book, across all three transport mechanisms, Table 5.8 is provided to summarize the various dimensionless groups employed. The application of dimensional analysis, coupled with proper experience and careful consideration, has proven to be a valuable tool in many areas of engineering.

---

[19]For laminar flow this relationship is only appropriate for negative pressure gradients, but in turbulent flow, conditions are less sensitive to the effect of pressure gradients and the relationship remains approximately valid [6].

**Table 5.8** Dimensionless groups and numbers used in this book

| Group, dimensionless nr. | Definition   Equation nr. | Use |
|---|---|---|
| Thermal Biot | $\mathrm{Bi_T} \equiv \overline{h}A/(\lambda_s A/L)$   (2.47) | Conduction/convection |
| Thermal Fourier | $\mathrm{Fo_T} \equiv \theta\alpha/L^2$   (2.95) | Conduction |
| Reynolds | $\mathrm{Re} \equiv \rho_f\langle w\rangle L/\mu$   (3.2) | General fluid flows |
| Fanning friction factor | $f \equiv \tau_w/\left(\frac{1}{2}\rho_f\langle u\rangle^2\right)$   (3.35) | Internal flows |
| Skin friction coefficient | $C_f \equiv \tau/\left(\frac{1}{2}\rho \mathbf{w}_\infty^2\right)$   (3.122) | External flows |
| Grid Peclet | $\mathrm{Pe} = \rho_f u\Delta x/\mu$   (Note 63, Chap. 3) | Num. discretization |
| Number of Transfer Units | $\mathrm{NTU} \equiv UPL/\left(\dot{m}c_p\right)_\mathrm{H}$   (4.19) | HEX performance |
| Prandtl | $\mathrm{Pr} \equiv \mu c_{pf}/\lambda_f = \nu/\alpha_f$   (4.50) | Heat/fluid flow |
| Turbulent Prandtl | $\mathrm{Pr_t} \equiv \varepsilon_\nu/\varepsilon_\alpha$   (4.51) | Turbulent flows |
| Peclet | $\mathrm{Pe}_L \equiv \mathbf{w}_\infty L/\alpha_f$   (4.52) | Convective/molecular h.t. |
| Eckert | $\mathrm{Ec} \equiv \mathbf{w}_\infty{}^2/\left(c_{pf}\Delta T\right)$   (4.53) | Convective heat transfer |
| Nusselt | $\mathrm{Nu}_L \equiv h_T L/\lambda_f$   (4.56) | Convective heat transfer |
| Thermal Grashof | $\mathrm{Gr}_{LT} \equiv g\beta\left(T_p - T_\infty\right)L/\mathbf{w}_\infty^2$   (4.70) | Natural convection h.t. |
| Rayleigh | $\mathrm{Ra}_L \equiv$ $g\beta_T\left(T_p - T_\infty\right)L^3/(\nu\alpha_f)$   (4.74) | Natural convection h.t. |
| Mass Biot | $\mathrm{Bi_M} \equiv \overline{h}_M L/D_s$   (5.44) | Diffusion/convection |
| Mass Fourier | $\mathrm{Fo_M} \equiv \theta D_s/L^2$   (5.45) | Diffusion |
| Thermal Richardson | $\mathrm{Ri}_{LT} \equiv \mathrm{Gr}_{LT}/\mathrm{Re}_L^2$   (5.59) | Mixed convection h.t. |
| Mass Richardson | $\mathrm{Ri}_{LM} \equiv \mathrm{Gr}_{LM}/\mathrm{Re}_L^2$   (5.60) | Mixed convection m.t. |
| Mass Grashof | $\mathrm{Gr}_{LM} \equiv$ $g\beta_M\left(T_p - T_\infty\right)L/\mathbf{w}_\infty^2$   (5.61) | Natural convection m.t. |
| Pomerantsev | $\mathrm{Po}_L \equiv \dot{e}'''L^2/(\lambda_f\Delta T)$   (5.62) | Source term effect in h.t. |
| Schmidt | $\mathrm{Sc} \equiv \nu/D_f$   (5.63) | Flow with mass transfer |
| Damkohler | $\mathrm{Da}_L \equiv \dot{m}_1'''L/(\mathbf{w}_\infty\Delta\rho_1)$   (5.64) | Source term effect in m.t. |
| Sherwood | $\mathrm{Sh} \equiv \hat{j}_{1p}x/(\Delta cD_f)$   (5.73) | Convective mass transfer |
| Thermal Stanton | $\mathrm{St_{T}}_x \equiv h_T/\left(\rho_f c_{pf}|\mathbf{w}_\infty|\right)$   (5.83) | Convective heat transfer |
| Mass Stanton | $\mathrm{St_{T}}_x \equiv h_M/|\mathbf{w}_\infty|$   (5.84) | Convective mass transfer |
| Lewis | $\mathrm{Le} \equiv \alpha/D_f$   (5.88) | Convective h. and m. t. |

## 5.5   Numerical Solution of Mass Transfer

We exploited the opportunity already to lay down the numerical solution bases, with Sect. 4.6 (p. 192), for the *macroscopic transport of any scalar of interest*. Let us drop for the time being the subscript for mass diffusivity $D$, for sake of simplicity.

### 5.5.1   Steady-State Mass Convection

We just need to restate, for the geometry in Fig. 4.18 and recalling the Fundamental Equation of mass transfer by diffusion in 1-D, in terms of *creation* of molar concentration of species $i$ and its *variable property and steady-state terms* version Eq. (5.36):

$$\frac{d}{dx}(uc_i) = \frac{d}{dx}\left(D\frac{dc_i}{dx}\right) + \frac{\dot{m}_i'''}{M_i}\Delta x_{CV} \tag{5.90}$$

Let us now drop, for sake of simplicity, the subscript $i$. Integrating Eq. (5.90) over the $CV$ of $P$ in Fig. 4.18, we get

$$(cu)_e - (cu)_w = D\frac{dc}{dx}\bigg|_e - D\frac{dc}{dx}\bigg|_w + \frac{\dot{m}'''}{M}\Delta x_{CV} \tag{5.91}$$

To simplify the evaluation of the variable at the interface points, similarly than with Eq. (3.119), let us assume that $w$ and $e$ lay midway between neighboring grid points:

$$c_e = \frac{1}{2}(c_E + T_P) \quad and \quad c_w = \frac{1}{2}(c_P + c_W) \tag{5.92}$$

then we can write, using again the definition of Eq. (2.106):

$$\frac{1}{2}u_e(c_E + c_P) - \frac{1}{2}u_w(c_P + c_W) = D_e\frac{c_E - c_P}{\Delta x_e} - D_w\frac{c_P - c_W}{\Delta x_w} + \frac{\dot{m}'''}{M_i}\Delta x_{CV} \tag{5.93}$$

In order to employ the matrix representation already envisaged in Sects. 2.5 (p. 55), 3.6 (p. 120), and 4.6 (p. 192), it is useful to cast Eq. (5.93) again in the following form:

$$\boxed{a_P c_P + a_E c_E + a_W c_W = b} \tag{5.94}$$

where

$$a_E = \frac{1}{2}u_e - \frac{D_e}{\Delta x_e} \tag{5.95a}$$

$$a_W = -\frac{1}{2}u_w - \frac{D_w}{\Delta x_w} \tag{5.95b}$$

$$a_P = \frac{1}{2}u_e - \frac{1}{2}u_w + \frac{D_e}{\Delta x_e} + \frac{D_w}{\Delta x_w} = -(a_E + a_W) + u_e - u_w \qquad (5.95c)$$

$$b = \frac{\dot{m}_i'''}{M_i}\Delta x_{CV} \qquad (5.95d)$$

As usual, while forming the final algebraic system, a general form like Eq. (5.94) can be cast for any internal grid point. For the Equations involving the boundary grid points, the considerations reported in Sect. 2.5.2.2 (p. 58) apply, along these made in Sects. 2.5.2.3 and 2.5.2.2 for the solution of the resulting algebraic system and the related treatment of nonlinearities.

Discretization of the Fundamental Equation of mass diffusion in 2-D is also straightforward, with the ideas and principles that were recalled in Sect. 3.6.2 (p. 125).

## 5.5.2 Unsteady-State Mass Convection

Turning to the unsteady-state version of the fundamental Equation of mass transfer in 1-D, for sake of simplicity in terms of *constant creation* of molar concentration of species $i$, by including term ① (the transient convective flux) of Eq.(5.35). Following the same line of thought leading to Eq. (4.124), starting from the grid point values of $T^0$ at time $\theta$, we will evaluate the "new" values $c$ at the time $\theta + \Delta\theta$. Therefore, integrating Eq. (5.35) over the $CV$ of $P$ (Fig. 5.8), we get

$$\int_w^e \int_\theta^{\theta+\Delta\theta} \left(\frac{\partial c}{\partial\theta} + u\frac{\partial c}{\partial x}\right) dx d\theta = \int_\theta^{\theta+\Delta\theta} \left[\left(D\frac{\partial c}{\partial x}\right)_e - \left(D\frac{\partial c}{\partial x}\right)_w\right] d\theta + \frac{\dot{m}_i'''}{M_i}\Delta x_{CV} \quad (5.96)$$

Again, nothing new needs to be added, with respect to the considerations exploited in Sect. 2.5.3 (p. 18), so that Eq. (5.96) becomes

$$\left[\frac{1}{2}u_e\left(c_E + c_P\right) - \frac{1}{2}u_w\left(c_P + c_W\right)\right]\Delta\theta + \Delta x_{CV}(c_P - c_P^0) =$$
$$\left(D_e\frac{c_E - c_P}{\Delta x_e} - D_w\frac{c_P - c_W}{\Delta x_w}\right)\Delta\theta + \frac{\dot{m}_i'''}{M_i}\Delta x_{CV} \qquad (5.97)$$

**Fig. 5.8** W and E points surrounding the P, for the 1-D problem, and the CV faces locations w and e

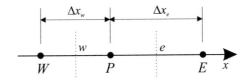

By using again a fully implicit scheme, we come up with

$$a_P c_P + a_E c_E + a_W c_W = b \tag{5.98}$$

where

$$a_E = \frac{1}{2}(\rho u)_e - \frac{D_e}{\Delta x_e} \tag{5.99a}$$

$$a_W = -\frac{1}{2}(\rho u)_w - \frac{D_w}{\Delta x_w} \tag{5.99b}$$

$$a_P^0 = -\frac{\Delta x_{CV}}{\Delta \theta} \tag{5.99c}$$

$$a_P = -\left(a_E + a_W + a_P^0\right) \tag{5.99d}$$

$$b = a_P^0 \left(c_P^0 + \frac{\dot{m}_i'''}{M_i}\right) \tag{5.99e}$$

In case of variable properties and source/sink term case, nothing new needs to be added with respect to the development presented at Chaps. 3 and 4.

## 5.6  Further Reading

- Phases and mixtures. Atkins, P.W., de Paula, J.: Physical Chemistry, Ninth Edition. Chapter 5. Oxford University Press, Oxford (2010)
- Derivation of the mass transfer equation; Dimensional analysis of the mass transfer equation. Belfiore, L.A.: Transport phenomena for Chemical Reactor Design. Chapters 9 and 10. John Wiley & Sons, Hoboken (2003)
- Dimensionless numbers in mass transfer and their meaning. Kuneš, J.: Dimensionless Physical Quantities in Science and Engineering. Elsevier, London (2012)
- Combined modes and process applications. Jaluria, Y., Torrance, K. E.: Computational Heat Transfer. Chapter 8. Hemisphere Publishing, Washington (1986)
- Unit operations. Geankoplis, C.: Transport Processes and Separation Process Principles. Chapters 8–14. Prentice Hall Press, Uppe Saddle River (2003)
- A literature review on conjugate heat and mass transfer. Caccavale, P., De Bonis, M.V., Ruocco, G.: Conjugate Heat and Mass Transfer in Drying: a Modeling Review. Journal of Food Engineering (2016) doi: 10.1016/j.jfoodeng.2015.08.031

- Operation and scale-up in bioprocesses. Shuler, M.L., Kargi, F.: Bioprocess Engineering. Chapters 9–11. Prentice Hall Press, Upper Saddle River (2002)
- Reaction mechanisms and theory of elementary reaction rates. Gardiner, W.C.: Rates and Mechanisms of Chemical Reactions. W.A. Benjamin, Menlo Park (1972)

# References

1. Faghri, A., Zhang, Y.: Transport Phenomena in Multiphase Systems. Academic Press, Burlington (2006)
2. Liley, P.E., Thomson, G.H., Friend, D.G., Daubert, T.E., Buck, E.: Physical and Chemical Data. In: Green, D.W., Maloney, J.O. (eds.) Perry's Chemical Engineers' Handbook, Chapter 2 (1997)
3. Irvine Jr., T.F.: Thermophysical Properties. In: Rohsenow, W.M., Hartnett, J.P., Cho, Y.I. (eds.) Handbook of Heat Transfer. McGraw-Hill, New York (1998)
4. Bejan, A.: Heat Transfer. Wiley, New York (1993)
5. Bird, R.B., Stewart, W.E., Lightfoot, E.N.: Transport Phenomena. Wiley, New York (2002)
6. Bergman, T.L., Incropera, F.P., Lavine, A.: Fundamentals of Heat and Mass Transfer. Wiley, New York (2011)

# Chapter 6
# Modeling Examples of PCB Processes

**Abstract** Many processes involve simultaneous transport phenomena in a variety of environments. The intertwined effects of gradients of temperature, velocity, and species concentration in the considered media are impossible to study by analytical methods and prohibitive to attack with experimental procedures. That is why numerical computations of such complex processes, or virtualization, come into play. Being the CFD our tool of choice to integrate the ensemble of governing PDEs, supplemented with proper PCB notations, one can enforce the virtualization of transfer phenomena to extract all of the information needed without recurring to systematic experimental assessment, provided that modeling has been properly validated with the corresponding data. In this last chapter, our attention is devoted to actual modeling: that is, the choice of proper governing Equations (with related initial and boundary conditions and ancillary statements, where applicable) that will be integrated and that will yield for all sort of graphical restitution of results.

## 6.1 Boundary Layer Simulation to Evaluate Sailboat Performance

Not a true multiphysics application, the pure transfer of momentum is the basis for intertwined computations. In this first modeling example, we will be briefly describing the flow of atmospheric air over an horizontal surface, creating a vast boundary layer flow, as it is found in many environmental situations. Boat sailing is one of these. Its simulation becomes readily complex when realism is brought into the calculations; nevertheless, some simplified modeling may be useful to summarize the work we have been preparing since Chap. 3.

A sailboat is a compound system that moves being immersed in two separated fluids: water and air. The two fluids are in relative motion to boat's hull and upper works, producing forces leading the skipper to appropriately evaluate course and adopt related maneuvers. As recalled by Sect. 3.1.2 (p. 70), the fluid is characterized

---

The original version of this chapter was revised: Belated corrections have been incorporated. The correction to this chapter is available at https://doi.org/10.1007/978-3-319-66822-2_7

© Springer International Publishing AG 2018
G. Ruocco, *Introduction to Transport Phenomena Modeling*,
https://doi.org/10.1007/978-3-319-66822-2_6

by the ability to flow and by its inherent viscosity, producing fluid and skin friction. When in contact with solid/fluid surfaces, the flow will deform: the water in contact with the hull, the airflow with the sails, the overall wind with the sea surface. When studying the principles of fluid deformation, we realized that this occurs also at a distance from the limiting surfaces.

Once devised properly, a numerical solution of momentum transfer can be employed to study the behavior of a sailboat during navigation. For the combination of fluids at stake, properties and relative motion states are different: water, with density $\rho$ of 3 OoM greater than air, can assume a maximum speed $\mathbf{w}$, relative to the hull, of a few meters per second. Instead, air can be even threefold faster relatively to the upper works, specially during close-hauled (upwind) sailing. These considerations dictate that the boat usually proceeds into water in laminar regime, whereas air flow past the upper works and over the sea largely in a turbulent regime. It is this last flow which is at stake in this model.

Let us consider a "Mini 6.50" sailboat (used in solo transatlantic races) as depicted in Fig. 6.1, with a mast and a maximum beam 12 and 3 m long, respectively, sailing close-hauled with a suitable heel angle (mast inclination with respect to the vertical) of 30°, which is typical of this sailing trim. As mentioned above, the largest relative aerodynamic force is detected on the main sail just aft of the mast (see Fig. 6.1, left).

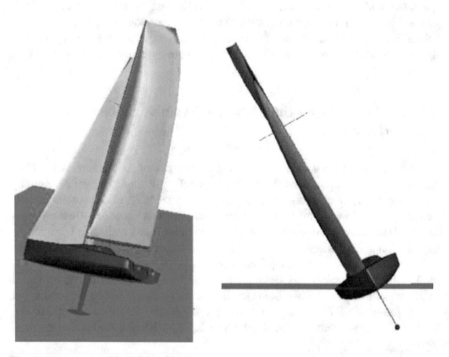

**Fig. 6.1** Two views of the Mini 6.50 navigating upwind. Colors represent a typical aerodynamic force distribution on the sails' sides, with red the highest value. Courtesy Dr. P. Caccavale, www. fluere.it (2016)

Then, the following *assumptions* are made:

- the boat sails without any external influence, other than the inherent wind, in a wide boundary layer;
- the wind flow is steady and moderately turbulent; therefore, the average velocity and pressure are at stake;
- the air flow is incompressible with constant properties;
- no-slip is enforced at every solid surface (the wind with the sails and the sea surface);
- the wind blows from a considerable distance; therefore, the boundary layer the boat is immersed in is fully developed (see Sect. 3.5.1). Let us assume that a number of measurements have been performed, in various heights from the sea surface, to quantify this boundary layer. For our scopes, a power-law velocity profile is mostly suited:

$$\mathbf{w}_\infty(z) \equiv |\mathbf{w}_0| \left(\frac{z}{z_0}\right)^{1/n} \tag{6.1}$$

where $|\mathbf{w}_0|$ is a velocity reference, such as the one measured by the wind gauge located on the boat mast, at a height $z_0$. In this case, the following values have been adopted:

$$|\mathbf{w}_0| = 6.0 \text{ kn}, \quad z_0 = 9.0 \text{ m}, \quad n = 5 \tag{6.2}$$

The resulting velocity profile is shown in Fig. 6.2.

- along its course, the boat keeps an upwind angle $\phi$ of 35°, which ensures a steady navigation speed relative to the water of $|\mathbf{w}_b| = 5.0$ kn

For the navigation field $CV$ reported in Fig. 6.3, the entering wind along $x$ has $u_\infty(z) \equiv -|\mathbf{w}_\infty(z)|$, and if $(u, v, w)$ are the absolute wind velocity components, the velocity components relative to the sails are $(u^*, v^*, w^*)$, with $u^* = u - |\mathbf{w}_b| \cos \phi$, $v^* = v + |\mathbf{w}_b| \sin \phi$ and $w^* = w$. The local relative wind intensity is therefore

$$|\mathbf{w}| \equiv \sqrt{u^{*2} + v^{*2} + w^{*2}}$$

**Fig. 6.2** The fully developed velocity profile of Eq. (6.1), with the values of Eq. (6.2), entering the $CV$ at stake

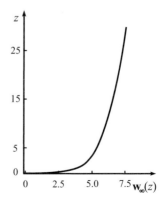

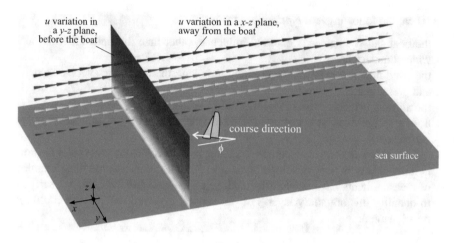

**Fig. 6.3** A view of the $CV$, with the boat (only the sails are reported, for the sake of simplicity) sailing close-hauled with course angle $\phi$. The $CV$ is 150 and 100 m wide along $x$ and $y$, respectively, while it is 30 m thick. The computed wind intensity distribution is represented with two colored renderings: *arrows* (in a generic $x - z$ plane) and *slice* (in a generic $y - z$ plane). Colors (from dark red to blue) represent the relative wind intensity values due to the inherent wind boundary layer, resulting from applying the fully developed velocity profile of Fig. 6.2

With reference to the previous statements, a standard steady-state RANS $k - \varepsilon$ model yielding for the average velocity components and average pressure in the subject $CV$ is enforced, by recalling the governing Equations set (Fig. 6.18):

$$\frac{\partial (w_i)}{\partial x_i} = 0 \qquad \text{(3.87 revisited)}$$

$$\rho_{\mathrm{f}} \frac{\partial}{\partial x_j} \left( w_i w_j \right) = \frac{\partial}{\partial x_j} \left[ (\mu + \mu_{\mathrm{t}}) \frac{\partial w_i}{\partial x_j} \right] - \frac{\partial p}{\partial x_i} \qquad \text{(3.102 revisited)}$$

$$\frac{\partial \left( w_j k \right)}{\partial x_j} = \frac{1}{\rho_{\mathrm{f}}} \frac{\partial}{\partial x_j} \left[ \left( \mu + \frac{\mu_{\mathrm{t}}}{\sigma_k} \right) \frac{\partial k}{\partial x_j} \right] - \varepsilon + \frac{1}{\rho_{\mathrm{f}}} S_k \qquad \text{(3.96 revisited)}$$

$$\frac{\partial \left( w_j \varepsilon \right)}{\partial x_j} = \frac{1}{\rho_{\mathrm{f}}} \frac{\partial}{\partial x_j} \left[ \left( \mu + \frac{\mu_{\mathrm{t}}}{\sigma_\varepsilon} \right) \frac{\partial \varepsilon}{\partial x_j} \right] + \frac{1}{\rho_{\mathrm{f}}} S_\varepsilon \qquad \text{(3.98 revisited)}$$

All definitions of Eqs. (3.99, 3.101–3.103) apply. Standard logarithmic "wall functions" for velocity also apply where necessary, depending on the CFD code employed.

The *boundary conditions* are particularly simple, in this case:

- at sea surface and at sails, the no-slip condition dictates that

$$\mathbf{n} \cdot \mathbf{w} = 0 \qquad (6.3)$$

with **n** denoting, as usual, the generic normal version for each delimiting surface
- at the leftmost $y - z$ inlet plane (see Fig. 6.3), the fully developed profile of Eq. (6.1) is applied
- at the remaining 4 $CV$ outlets, the BC of zero pressure and no backflow are assumed

The system of Eqs. (3.89, 3.104, 3.98, 3.100) is discretized over the subject navigation field $CV$, turning into six large systems of $n$ algebraic equations (with $n$ the number of discretized grid points, leading to FV or FE formulations). Each system ensemble (with Note 49 of Chap. 2 in mind) would bring the solution for $p$, $k$, $\varepsilon$ and each $w_i$ component. This very large equation ensemble is fed into an available *algebraic solvers*, following the chosen direct or iterative solution strategy, as anticipated in Sect. 2.5.4 (p. 62).

Upon solution of the flow field, the wind distribution can be represented as depicted by Fig. 6.3. Now, an object that protrudes in a boundary layer produces a perturbation of the flow field downwind. This is precisely the situation depicted in Fig. 6.4, where a view of the solution of the relative flow field is reported, by considering four vertical slices normal to the wind (from ① to ④), whose color distribution represents wind intensity in each considered location in the $CV$ domain.

Slice ① displays the undisturbed windward boundary layer, corresponding to the situation already depicted in Fig. 6.3. Next, with slice ② we see that the wind is strongly slowed down (from an intensity of ≈7.5 to almost 0 kn) by friction, downwind to the sails. Slices ③ (at about 30 m downwind) and ④ (at about 60 m downwind) evidence the formation of a "disturbance tunnel," with bell-shaped sections colored in yellow/green, attenuating and smoothing out with distance. This effect motivates

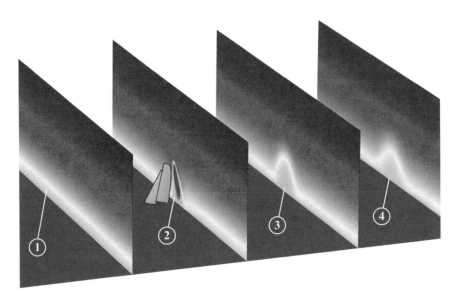

**Fig. 6.4** Numerical results of wind intensity presented by 4 different slices labeled from ① to ④, normal to wind direction

**Fig. 6.5** An indicative
turbulence dome extending
aft from the forestay

forestay

for the need, when sailing downwind to a boat, to keep well away such invisible disturbance, which gives its effect even from a considerable distance upwind. Indeed, this occurrence can considerably hinder the boat's otherwise normal course, by decreasing her navigation speed and even worsening (increasing) her close-haul angle $\phi$.

Interacting with the sails, then, the air is strongly slowed down and disorder is induced in the flow past them. In Fig. 6.5, three isoturbulence surfaces (each featuring a single value of turbulence intensity) are seen extending aft starting from the forestay, which is the most forward sail edge directly exposed to the incoming wind. Nested one inside another, with increasing values of intensity inward, these isosurfaces highlight shape and downwind extension of a strong "turbulence dome."

The wind-sail aerodynamic interaction alters very noticeably the wind direction and intensity, otherwise uniform along $x$ and with no $y$-component (see coordinate reference frame in Fig. 6.3). Figures 6.6 and 6.7 illustrate the local effects that are created in every portion of the combined rig (consisting in jib and mainsail). Here, some horizontal ribbons are adopted that help visualize a vertical airflow "blade." In Fig. 6.6, this vertical flow is seen from far-aft. The wind ribbons, coming from the left, have their undisturbed red-orange intensity, but, as they approach the rig leading edge, are pushed away from the sails slowing down, then resuming their pristine intensity past the boat. The same situation is shown in Fig. 6.7, from a far-bow perspective, where is the flow ribbon deformation is again evident, specially in the upper half of the rig: there, the deceleration effect from the undisturbed boundary layer intensity (ribbon color temporarily shifting to blue when passing behind the jib) is stronger, due to the minimum distance between the ribbons and the sail convexity. The wind bypassing the sails behaves in the same overall way as producing a lift at an airplane's cambered airfoil, or a downforce at a Formula 1 car's rear spoiler (Fig. 6.8): the displacement of the ribbons past the sail, to a vertical plane located away downwind, evidences the *variation in intensity and direction of the deflected airflow momentum*, which must be balanced by a resulting lift force, applied to the airfoil or sail geometrical center. While sailing, this force amounts to the intended thrust, properly managed by the underwater keel, that drives the boat along her course.

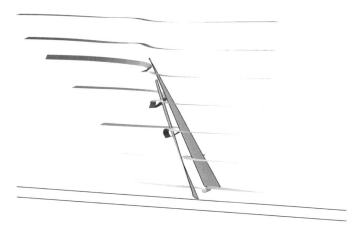

**Fig. 6.6**  Wind ribbons through and past the sails, as seen from far-aft

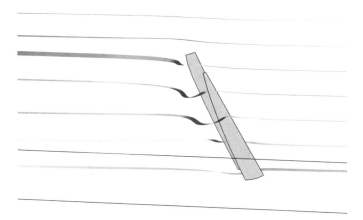

**Fig. 6.7**  Wind ribbons through and past the sails, as seen from far-bow

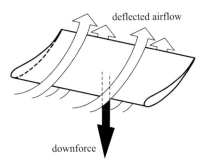

**Fig. 6.8**  A simplified vision of the aerodynamic downforce from an airfoil spoiler due to airflow deflection. More complex than the scheme reported here, the fluid dynamic interaction between a fluid flow and an airfoil also features transversal components and the creation of vortical structures in the flow field past the structure

Physical replicas of these computational ribbons are usually provided on actual rigs to each sail leech, called the telltales, which would then behave exactly as seen in this virtualization, giving useful indications to the skipper for his/her decisions on course maneuvers. All in all, this kind of simulations can become a reliable tool for sail designers to optimize their shape and interaction.

## 6.2   Cooling of Cylindrical Substrates in a Complex Flow Field by a Conjugate Formulation

One important feature of multiphysics modeling is the ability to perform computations across more phases, without the limitations of a separating interface. In this second modeling example, we will be examining conduction heat transfer (in a solid s) due to localized convection (in a fluid f). Substrates of any kind and shape can be processed by applying or subtracting heat for a number of purposes. When dealing with problems such as exposure to working fluids, insulation, and heat flux application, conduction comes into play. For example, freezing times of substrate protrusions by forced air convection can be calculated for internal uniformity monitoring, with the local and transient temperature distribution in the substrate depending on the intensity and flow patterns of the applied convection.

Air jet impingement (JI) is an important option in enhancement and control. A localized air draft by JI yields an increment of the heat transfer to/from impinged items, even by using only low turbulence flows. But the same procedure can be employed to monitor and mitigate otherwise uncontrolled processing. This paves the way to improved convection processes with respect to traditional bulk air drafts.

When a conductive protrusion is exposed to fluid flow as the heat transfer medium, a modern modeling procedure is to prepare for a conjugate heat transfer analysis, as anticipated in Sect. 4.5.2 (p. 179) that computes the temperature distribution simultaneously in both solid and fluid phases, in order to determine the solution for temperature with no need for empirical assumptions of the convective heat transfer coefficient at the interface. Upon calculation, the resulting flow field and its relationship with the temperature distribution within the protrusion can be deduced; depending on the adopted operating conditions, the most unfavored point for the cooling process, or slowest cooling zone (SCZ), can be localized elsewhere than with traditional cooling calculations. This is important to evaluate the energy expenditure of the process, and to avoid under- or overtreatments.

Let us consider a cold air JI flow directed, from above, over a long cylindrical biosubstrate protrusion as depicted in Fig. 6.9.[1] The flow interacts with the solid

---

[1] As performed by Dirita, C., De Bonis, M.V., Ruocco, G.: Analysis of food cooling by jet impingement, including inherent conduction (2007) https://doi.org/10.1016/j.jfoodeng.2006.10.002..

**Fig. 6.9** A cooling air jet onto a cylindrical protrusion: nomenclature of flow regions. Reprinted from Journal of Food Engineering, Vol 81, Carmela Dirita, Maria Valeria De Bonis, Gianpaolo Ruocco, Analysis of food cooling by jet impingement, including inherent conduction, 12–20, Copyright (2007), with permission from Elsevier

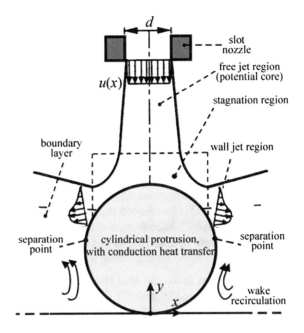

forming a complex flow field around the object, with many flow regions. The free jet region (potential core) is perturbed at first by becoming a stagnation region above the protrusion, where the highest heat transfer rate is expected; then, a wall jet develops, with decreasing heat transfer rate, until the flow separates past the separation point, even for relatively small values of the Reynolds number. The wall jet can be considered a particular case of deformed boundary layer, bearing more complexity with respect to the one examined in Sect. 3.5.2 (p. 113). The separation of flow drastically reduces the heat transfer, which eventually resumes downwind, due to the residual effect of a wake, that is anticipated with respect to the classical flows across cylinders (as in Fig. 3.21). It is clear, therefore, that the formation of diverse flow regions will result in a complex mutual interaction between heat transfer modes (convection and conduction). This means that the local heat transfer rate is strongly affected, and with it the temperature distribution in the protrusion and its cooling uniformity.

Then, the following *assumptions* are made:

- JI at temperature $T_j$ discharges from a height $H = 12$ cm into a process space, limited by two horizontal plates and two lateral exhausts sufficiently placed apart
- the impinged protrusion, with diameter $D = 3.5$ cm, is a isotropic and homogeneous biomaterial with the following thermal properties: $\rho_s = 1100$ kg/m$^3$, $c_{ps} = 3390$ J/kgK and $\lambda_s = 0.55$ W/mK

- the average air velocity $\langle w \rangle$ at the jet nozzle is such that JI flow is fully developed turbulent, based on

$$\mathrm{Re} \equiv \frac{\rho_f \langle w \rangle d}{\mu} \approx 23,000 \qquad \text{(3.2 revisited)}$$

with $d = 30$ mm the slot nozzle width
- the protrusion cylinder is infinite along $z$, so that the solution can be computed on the $x - y$ plane, only
- the resulting flow is therefore two-dimensional and has temperature-dependent properties but is incompressible (negligible pressure work and kinetic energy)
- the viscous heat dissipation can be neglected, and no body force can be accounted for (unless the JI is terminated and a natural convection cooling is carried out)
- Dirichlet stationary boundary conditions are applied for temperature $T$ at limiting surfaces (nozzle wall and upper and lower plates)
- no-slip is also enforced at all solid surfaces

With reference to the previous statements, a standard steady-state RANS model of choice (as reported in Sect. 6.1 (p. 235), for example) and energy Equations are enforced yielding the average velocity components and pressure in the fluid phase, and average temperature in both phases. To this end, the governing Equations are recalled:

$$\frac{\partial \overline{T_f}}{\partial \theta} + \frac{\partial}{\partial x_j}\left(\overline{T_f}\overline{w}_j\right) = \frac{\partial}{\partial x_j}\left((\alpha_f + \varepsilon_\alpha)\frac{\partial \overline{T_f}}{\partial x_j}\right), \text{ in the fluid phase f} \qquad (4.47)$$

$$\frac{\partial T_s}{\partial \theta} = \frac{\partial}{\partial x_j}\left(\alpha_s \frac{\partial T_s}{\partial x_j}\right), \text{ in the solid phase s} \qquad \text{(2.22 revisited)}$$

The *initial and boundary conditions* are as follows:

- the substrate is initially in thermal equilibrium with the quiescent ambient air, $w_i = 0$ so that $T_s = \overline{T_f} = T_0$
- a proper turbulent profile is applied at nozzle inlet, such as the one suggested by

$$\overline{w_2}(x_1) = \left(1 - \frac{x_1}{d/2}\right)^{1/7}\overline{w_{2\mathrm{max}}} \qquad \text{(3.32 revisited)}$$

with $\overline{w_{2\mathrm{max}}}$ determined by the inherent Re number, Eq. (3.2) revisited above
- at the limiting upper and lower plates, the usual no-slip condition Eq. (6.3) and "wall functions" apply (as mentioned in the previous modeling example), whereas the temperature is maintained at the given environmental limiting value $T_\infty$
- at the lateral exits, far enough from the protrusion, the conditions of zero pressure and no backflow are assumed
- at the protrusion's exposed surface, beside the no-slip condition and the "wall functions," the following conjugate heat transfer condition is applied

$$\overline{T_f} = T_s$$

The system of chosen RANS supplemented with Eqs. (4.47, 2.22) is discretized over the subject fluid and solid phases $CV$ following the development of Sect. 4.6.3 (p. 192), turning into six large systems of $n$ algebraic equations (with $n$ the number of discretized grid points), plus another one for a single temperature $T_{conj}$ variable (in the spirit of the conjugate formulation). Therefore, beside the solution for the distribution for $w_i$ and $p$, each system ensemble (with Note 49 of Chap. 2 again in mind) would bring the solution for $T_{conj}$, as well. This very large system is fed into an available *algebraic solver*, following the chosen solution strategy, as anticipated in Sect. 2.5.4 (p. 62).

Following the same considerations made already in Sect. 2.5.5 (p. 64), the *gridding strategy* must allow for grid point increment in the zones where greatest velocity and temperature gradients are expected; dynamically changing grids are also used in sophisticated simulations. Based on the convergence criteria presented in Sect. 3.6.6.3 (p. 140), the adopted computational procedure must include a sufficient grid-independency "descent."

Upon solution of the flow and temperature fields, the velocity streamlines and the temperature distribution in the working fluid are represented in Fig. 6.10a. Recalling Fig. 6.9, the characteristic stagnation region, wall jet region, separation point, and wake recirculation are perfectly evident in Fig. 6.10b. As a consequence of this exposure to fluid flow, the temperature in the protrusion is shown in Fig. 6.11 to be readily perturbed: in less than two minutes, the heat is efficiently removed from the surface, cooled down to approximately $-22\,^{\circ}\mathrm{C}$, except in the bottom region of the substrate, due to the ineffective surface convection. Recalling the definition for

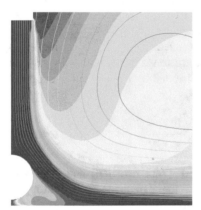
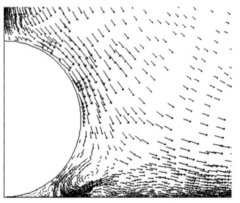

(a) A representation coupling coloured velocity streamlines and temperature contours in the fluid phase

(b) Close-up: velocity arrow distribution

**Fig. 6.10** Steady-state cooling flow field over and past the protrusion, as determined by the JI flow. In this case: $T_j = -35\,^{\circ}\mathrm{C}$, $T_0 = T_\infty = 0\,^{\circ}\mathrm{C}$; $\overline{w_{j\max}} \approx 12$ m/s. *Source* for Figure (b): see Fig. 6.9

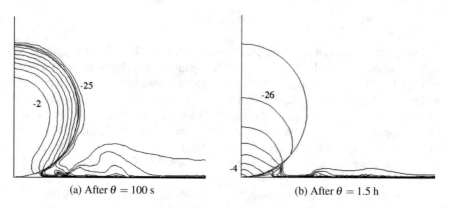

(a) After $\theta = 100$ s                                    (b) After $\theta = 1.5$ h

**Fig. 6.11** Distribution of conjugate temperature $T_{conj}$ (°C) in the protrusion and in the working fluid due to treatment of Fig. 6.10, with $\Delta T \approx 2.3$ °C between consecutive isotherms. *Source* see Fig. 6.9

the conduction flux in the protrusion

$$\dot{q}_{\lambda n} = -\lambda \frac{\partial T}{\partial \mathbf{n}} \tag{2.7}$$

this is seen to be initially directed radially outward except in the bottom region, Fig. 6.11a. With the cooling onset, due to the JI asymmetry on the exposed surface, $\dot{q}_{\lambda n}$ is directed only toward the protrusion's upper half, and the SCZ is found displaced laying on protrusion bottom, as shown in Fig. 6.11b.

The cooling effect is also displayed in Fig. 6.12 for several time frames by colored temperature contours, for the first minutes from beginning. It is again evident how the SCZ slowly moves downward during the treatment.

A variety of virtualized experiments can be carried over by using this model, as for pure natural convection. In this case, the protrusion is let to cool off in stagnant air, and the RANS adopted so far must be substituted by the unsteady version of Eq. (4.66)

$$\frac{\partial w_i}{\partial \theta} + \frac{\partial}{\partial x_j}\left(w_i w_j\right) = \frac{\partial}{\partial x_j}\left(v\frac{\partial w_i}{\partial x_j}\right) + \frac{g\beta(T_f - T_\infty)}{\rho_f} \tag{4.66 revisited}$$

which is supplemented again by the above Eq. (2.22 revisited) and the above Eq. (4.47), where the overline notations and the parameter $\varepsilon_\alpha$ are dropped. Figure 6.13 reports that the SCZ is again asymmetrical: now located midway between top and protrusion center, after less than 20 min.

Finally, the case at stake is instructive to learn on the difference between conventional (segregated) and conjugate heat transfer modeling and results. We anticipated these notations in Sect. 4.5.2 (p. 179). Let us consider the forced convection case seen earlier; for example, the segregated notation using the average convective heat

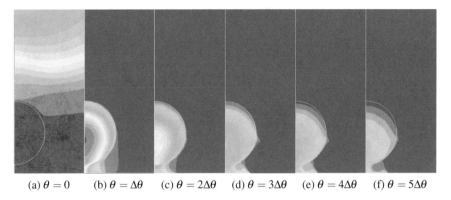

(a) $\theta = 0$     (b) $\theta = \Delta\theta$     (c) $\theta = 2\Delta\theta$     (d) $\theta = 3\Delta\theta$     (e) $\theta = 4\Delta\theta$     (f) $\theta = 5\Delta\theta$

**Fig. 6.12** Qualitative contours of conjugate temperature $T_{conj}$ in the protrusion and in the working fluid, due to the treatment of Fig. 6.10, for a variety of time frames proceeding by a small $\Delta\theta$

**Fig. 6.13** Distribution of conjugate temperature $T_{conj}$ (°C) in the protrusion and in the working fluid, due to natural convection after $t = 1000$ s, with $\Delta T \approx 2.9$ °C between consecutive isotherms. In this case: $T_0 = T_\infty = 0$ °C. *Source* see Fig. 6.9

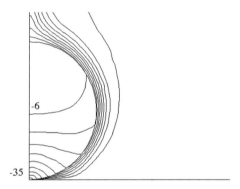

transfer coefficient

$$\overline{h}_T = -\frac{\lambda_f}{T_0 - T_\infty} \left.\frac{dT}{dy}\right|_{y=0^+} \tag{4.6a}$$

is clearly inadequate, as anticipated in Sect. 4.1.2.4 (p. 148), as the protrusion's surface is exposed to a variety of flow field effects and no heat transfer uniformity can be expected.

A rational approach to non-uniform convection heat transfer requires the use of a *local Nusselt number*

$$\mathrm{Nu}_x \equiv -\left.\frac{\partial T_f^*(x)}{\partial y^*}\right|_{y^*=0}$$

which extends on the local basis the concept presented already with Eq. (4.56b). To this end, we must adopt a proper local coordinate system $(\phi, y^*)$ on the protrusion's surface, as in Fig. 6.14, $\phi$ being the polar coordinate and $y^*$ a local coordinate normal

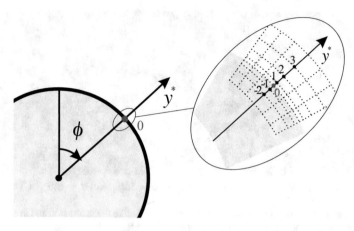

**Fig. 6.14** Local polar coordinates ($\phi, y^*$) on protrusion surface. In the inset, the locations of six grid points ($-2$–$3$) across the interface are reported, to interpolate the temperature derivative of Eq. (6.2). Also, the gridpoint intersection of the $y^*$ axis with the surface is highlighted in red. Note that a proper gridding (increment of grid line along $y^*$ in the vicinity of surface) is employed to resolve the temperature gradient across the surface, in the spirit of the conjugate formulation

to the surface, rendered non-dimensional with respect to $D/2$. So the local Nusselt number can become

$$\mathrm{Nu}_\phi \equiv \frac{h_\mathrm{T} D}{2\lambda_\mathrm{f}} = -\left.\frac{\partial T_\mathrm{f}^*(\phi)}{\partial y^*}\right|_{y^*=0}$$

while the *dimensionless temperature* $T_\mathrm{f}^*$ is based on a reference temperature difference

$$T_\mathrm{f}^* \equiv \frac{T_\mathrm{f} - T_\mathrm{j}}{T_0 - T_\mathrm{j}} \tag{6.4}$$

But we must acknowledge, following the motivation of a conjugate formulation, that a single-phase (segregated) computation of convection heat transfer, even in its locally varying version of Eq. (4.56) revisited above, is inadequate as well, as we cannot get rid of the underlying, ever-changing temperature of the protrusion. Having computed the whole temperature distribution $T_\mathrm{conj}$ in the entire $CV$, it is straightforward to use Eq. (6.4) to define a dimensionless $T_\mathrm{conj}$, so to come up with the *conjugate local Nusselt number*:

$$\mathrm{Nu}_\phi \equiv -\frac{\partial T_\mathrm{conj}^*(\phi)}{\partial y^*} \tag{6.5}$$

whose right-hand side can be computed via a $n$-point Lagrangian interpolation ($n = 6$, in this example), having two central points straddling the protrusion interface to fully capture the inner and outer temperature variations. This interpolating polynomial can be profitably chosen in cases, such as the one at stake, when

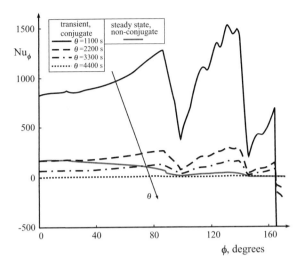

**Fig. 6.15** Local Nusselt number around the impinged protrusion, in transient/conjugate regime (with indication of process durations $\theta$) and steady-state/non-conjugate regime (in red). Data elaborated from Reference in Note 1

the independent variable (the grid point positions along $y$) is unevenly spaced [1]. Sample grid points for the evaluation of $T^*_{conj}$ and, ultimately, the computation of $Nu_\phi$ are indicated in the inset of Fig. 6.14. It is clear that the grid points pertaining to the segregated solution are solely 0–3.

Typical time progress of computed $Nu_\phi$ is reported in Fig. 6.15, at regular time intervals, confirming that the treatment at stake is eminently transient, specially in the first hour. At cooling start-up (for $\theta$ from 1100 to 2200 s), $Nu_\phi$ reduces sharply, three- or fourfold. Large discontinuities and decrements are evident in its progress around the exposed surface, due to separation and recirculation seen earlier. The separation point occurs steadily for $90° < \phi < 100°$. With the process onset (for $\theta$ from 3300 s on), these discontinuities are smoothed out as the protrusion is already cooled off, and much more heat transfer can be released no longer to the working fluid.

An important effect of the conjugate formulation is also seen toward the protrusion's bottom (after ca. 160°), as an inversion of the heat flux direction can be seen (the Nusselt number becomes negative): the protrusion is being heated up by the surrounding air.

Figure 6.15 also includes, for comparison, the progress of steady-state $Nu_\phi$ computed via Eq. (6.5) (as no correlation was available in this case). It is evident therefore that a segregated approach to solve convection heat transfer problem is completely inadequate, when a realistic effect of the temperature-varying conduction is allowed and the inherent transient behavior of the heat exchange is at stake.

## 6.3   Moist Substrates Subject to Intense Heat Generation

A second important feature of multiphysics modeling is to account for additional competing/contrasting mechanisms, such as microwaves (MW) or ultrasounds. Exposure to these mechanisms generally translates in strong source/sink terms, as we anticipated in Sect. 1.3.3 (p. 7). In this third modeling example, bulk thermal air convection is supplied to a substrate, when intense internal heat generation is added. Inherent phase-change of liquid water in the substrate is also accounted for. Processes such as this are carried out, for example, for *enhanced and controlled drying*.

MW-enhanced drying, a process nowadays somewhat common in a number of cases, features a physical mechanism different than the conventional convective drying. In the latter, water phase-change is induced by the thermal perturbation carried by the working forced air applied on the substrate's external surface and mediated within the substrate by heat conduction. In the former, the induced MW heating acts directly within the substrate instead, resulting by the interaction between the electromagnetic field and dipolar molecular species of substrate's liquid phase. This may be pure liquid water, solutions, or even salts: the friction produced by the dipoles rotation and by the migration of ionic species to regions of opposite charge generates heat, specially where the liquid phase is in relative excess. Drying by MW offers then several distinct benefits, including increasing throughput and higher energy efficiency, its intensity, and penetration depth varying with:

- electromagnetic frequency
- substrate's thermodynamic state, dictated by composition, and temperature-dependent dielectric properties
- substrate shape

MW treatments happen to be rather rapid, that is why a certain non-uniformity in the temperature distribution results in common biosubstrates: overheating loci may be attained within the substrate where the liquid phase is in relative excess, or even runaway temperatures (with substrate spoiling) may occur in some outer layer. For this reason, forced air convection may be contemporarily supplemented in the process inducing a desired superficial finish and mitigating the superficial temperature excess. In the end, in MW-enhanced drying all transfer phenomena are coupled and intertwined and need to be properly addressed in modeling, due to the coexistence of the following mechanisms:

- volumetric heat generation due to MW
- latent cooling due to phase-change
- heat conduction within the substrate
- evaporation of the free water from external surface and within the substrate
- diffusion of free water
- diffusion of water vapor produced by evaporation
- local variability of momentum transport on the exposed surface, driving

**Fig. 6.16** MW-enhanced drying: the variety of intertwined transport phenomena mechanism

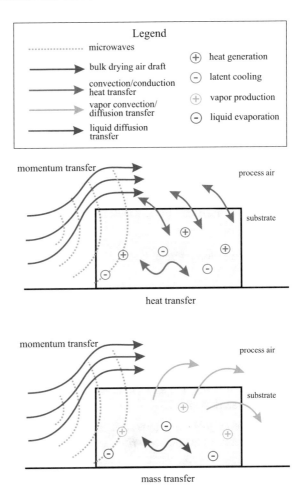

– local surface heating/ cooling
– local vapor transport and elimination from surface

All these mechanisms occur on a local basis. Figure 6.16 illustrates these combined occurrences. A multiphysics-integrated model will allow for the assessment of influence of the driving parameters (the combined convection-MW drying arrangement, the velocity of the working air and process duration) on the overall process (thermal and moisture distribution within the substrate). As with the previous modeling example in Sect. 6.2, a conjugate approach can be employed by solving simultaneously the heat and mass transfer in both substrate s and air a phases.

Let us consider a forced air draft in a long 2-D channel, flowing past a flat biosubstrate protrusion, simultaneously exposed to MW radiation produced by a nearby waveguide, as depicted in Fig. 6.17. The flow will interact with the substrate

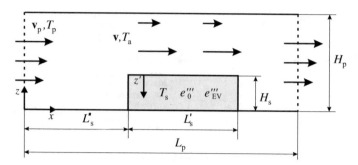

**Fig. 6.17** MW-enhanced drying of a flat substrate: geometry and operating parameters nomenclature. The $CV$ is composed by the substrate subdomain (gray) sitting in a air channel subdomain (white). Relative sizes are distorted for simplicity of presentation

forming a developing boundary layer along and past the protrusion, as shown earlier in Fig. 3.23.

The following *assumptions* are considered:

- the drying air, flowing in turbulent regime, is incompressible with temperature-dependent properties but with negligible viscous heat dissipation
- the substrate is homogeneous and isotropic; for the sake of simplicity, its dielectric properties are constant
- the thermal conductivity and mass diffusivity are considered to be temperature-dependent, while the specific heat and density are considered both temperature- and moisture-dependent

With reference to the previous statements, the governing Equations for energy, mass, and momentum transfer are written as follows:

*Energy transfer, in the sample*    First of all, an adequate formulation is needed to model the internal heat generation in the sample being dried. A complete approach requires the solution of the electromagnetic field distribution by integrating the *Maxwell's equations*.[2] Otherwise, in the presence of a thin substrate, it can be further assumed that plane MWs only are incident on the exposed upper surface of the substrate and that this radiation is uniform along the $x$-coordinate. With these assumptions in mind, an energy penetration depth can be calculated (even in the presence of multiple medium properties affecting absorption) by *Lambert's law*.[3] In this case, the governing Equation can be written as:

$$\rho_s c_{ps} \frac{\partial T_s}{\partial t} = \nabla \cdot (\lambda_s \nabla T_s) + \dot{e}_{MW}''' - \dot{e}_{EV}''' \qquad (2.22 \text{ revisited})$$

---

[2] As performed by De Bonis, M.V., Caccavale, P., Ruocco, G.: Convective control to microwave exposure of moist substrates. Part I: Model methodology. International Journal of Heat and Mass Transfer (2015) https://doi.org/10.1016/j.ijheatmasstransfer.2015.03.037.

[3] As attributed to the work by the German mathematician J.H. LAMBERT at the end of the eighteenth century.

where the usual symbols are employed and where

- $\dot{e}_{MW}'''$ $(W/m^3)$ is the volumetric power due to dielectric dissipative effect of exposure to MW (which depends on penetration depth or relative coordinate $z'$), given by:

$$\dot{e}_{MW}''' = \dot{e}_0''' f(z')$$

with $\dot{e}_0'''$ $(W/m^3)$ is the nominal MW volumetric power right from the waveguide, $z'$ the depth from the exposed surface, and $f(z')$ a dimensionless complex-valued energy distribution expression. Laying the specification of $f(z')$ outside the scope of this presentation, it is only recalled here that $f(z')$ changes considerably with thickness and substrate properties.[4] As an example, in Fig. 6.19 the value of $f(z')$ for two common foodstuffs at variable thickness is reported

- $\dot{e}_{EV}'''$ is the local cooling flux due to the latent heat of evaporation, given by

$$\dot{e}_{EV}''' = M \Delta h_{vap} K c_l$$

featuring

- $M$ as the *molecular weight of water* (kg/kmol)
- $\Delta h_{vap}$ as the *latent heat of evaporation of water* (kJ/kg)
- $K$ as the *rate of production of water vapor mass* or, more simply, *evaporation* (1/s)
- $c_l$ as the *local molar concentration of liquid water* (kmol/m³)

Incorporating the dependence on the local temperature $T$, a simple, first-order Arrhenius kinetics is often employed for the rate of production $K$[5]

*Energy transfer, in the working air*   To attain energy continuity across the sample's exposed surface, the governing Equation can be written as:

---

[4] $f(z')$ depends on media's *dielectric properties*, that dictate the transmission and reflection coefficients, and the attenuation and phase factors. A complete account on $f(z')$ for a variety of moist substrate can be found in the accompanying paper by Marra, F., De Bonis, M.V., Ruocco, G.: Combined microwaves and convection heating: A conjugate approach. Journal of Food Engineering (2010) https://doi.org/10.1016/j.jfoodeng.2009.09.012.

[5] After Swedish physical chemist S.A. ARRHENIUS, early twentieth century:

$$K = A \exp(E_a/RT_s)$$

with the pre-exponential factor $A = 925$ (1/s), $R$ (J/kmolK) the universal gas constant, and $E_a$ (J/mol) a *dehydration activation energy*, whose value depends on the heating physics employed. As an example, $E_a = 4 \times 10^4$ for air convection dehydration, while for a combined air convection/microwave dehydration $E_a = 3.1 \times 10^4$ (being the later mechanism less demanding in terms of energy budget for drying). This approach has been first exploited, in a conjugate heat and mass transfer framework, by De Bonis, M.V., Ruocco, G.: A generalized conjugate model for forced convection drying based on an evaporative kinetics. Journal of Food Engineering (2008) https://doi.org/10.1016/j.jfoodeng.2008.05.008.

$$\frac{\partial \overline{T}_a}{\partial \theta} + \frac{\partial}{\partial x_j}\left(\overline{T}_a \overline{w}_j\right) = \frac{\partial}{\partial x_j}\left((\alpha_f + \varepsilon_\alpha)\frac{\partial \overline{T}_a}{\partial x_j}\right) \qquad \text{(4.47 revisited)}$$

*Mass transfer, in the sample*    As we note that the energy Equation in the sample contains the *local molar concentration of liquid water* $c_l$ (kmol/m$^3$), to attain the problem closure a supplementing governing Equation must be written:

$$\frac{\partial c_l}{\partial \theta} = \nabla \cdot (D_{ws}\nabla c_l) - K c_l \qquad \text{(5.39 revisited)}$$

where $D_{ws}$ (m$^2$/s) is the diffusivity of water in the substrate. Due to evaporation, water vapor will be produced within the sample and at the exposed surface, as dictated by the following governing Equation:

$$\frac{\partial c_v}{\partial \theta} = \nabla \cdot (D_{ws}\nabla c_v) + K c_l \qquad \text{(5.39 revisited)}$$

where $c_v$ appears as the *local molar concentration of water vapor* (kmol/m$^3$). In the last two Equations, their source terms are given by the enforcement of the mass conservation of water, since a mole of evaporated liquid corresponds to a mole of vapor produced in the $CV$

*Mass transfer, in the working air*    Water vapor at the exposed surface will be depleted by the air draft:

$$\frac{Dc_v}{D\theta} = \nabla \cdot (D_{wa}\nabla c_v) \qquad \text{(5.38 revisited)}$$

where $D_{wa}$ is the diffusivity of water in the air

*Momentum transfer*    Problem closure is attained as soon as we recognize the energy Equation in the working air contains the air velocity **v**. If the channel Reynolds number

$$\text{Re}_{H_p} \equiv \frac{\rho_a \langle v \rangle H_p}{\mu} \qquad \text{(3.2 revisited)}$$

exceeds a certain velocity magnitude threshold, turbulent flow occurs, so that we need to invoke the following governing RANS Equations

$$\frac{\partial (\rho_a \overline{v}_i)}{\partial x_i} = 0 \qquad \text{(3.88}a\text{ revisited)}$$

$$\rho_a \frac{\partial \overline{v}_i}{\partial \theta} + \rho_a \frac{\partial}{\partial x_j}\left(\overline{v}_i \overline{v}_j\right) = \frac{\partial}{\partial x_j}\left[(\mu + \mu_t)\frac{\partial \overline{v}_i}{\partial x_j}\right] - \frac{\partial \overline{p}}{\partial x_i} + \overline{S}_i \qquad \text{(3.102 revisited)}$$

with $\overline{S}_x = 0, \overline{S}_y = \rho_a g(T_a - T_0)/T_0$, and

$$\frac{\partial (\rho_a k)}{\partial \theta} + \frac{\partial (\rho_a \overline{w}_j k)}{\partial x_j} = \frac{\partial}{\partial x_j} \left[ \left( \mu + \frac{\mu_t}{\sigma_k} \right) \frac{\partial k}{\partial x_j} \right] - \rho_a \varepsilon + S_k \qquad \text{(3.96 revisited)}$$

$$\frac{\partial (\rho_a \varepsilon)}{\partial \theta} + \frac{\partial (\rho_a \overline{w}_j \varepsilon)}{\partial x_j} = \frac{\partial}{\partial x_j} \left[ \left( \mu + \frac{\mu_t}{\sigma_\varepsilon} \right) \frac{\partial \varepsilon}{\partial x_j} \right] + S_\varepsilon \qquad \text{(3.98 revisited)}$$

where the ancillary definitions of the $k - \varepsilon$ turbulence model can be deduced by Sect. 3.3.7.5 (p. 104)

In moist substrate processing, that is a frequent subject in multiphysics modeling, it is customary to use a specially defined *dry/basis humidity* (moisture) content $X$.[6] The *initial and boundary conditions* are as follows:

- the sample is initially at equilibrium with the quiescent air, $w_i = 0$, and the cavity surfaces so that $T_s = T_a = T_0$, and has an initial moisture $c_{l0}$, while no vapor is initially assumed in the system, $c_{v0} = 0$
- with reference to Fig. 6.17, uniform conditions at the inlet (left) for velocity, pressure, turbulence parameters, and temperature are given: $\mathbf{v} = \mathbf{v}_p$, $p = p_p$, $k = k_p$, $\varepsilon = \varepsilon_p$, $T = T_p$
- at the outlet (right), the conditions of zero pressure and no backflow are assumed: $\partial \mathbf{v}/\partial x = 0$, $\partial k/\partial x = 0$, $\partial \varepsilon/\partial x = 0$, $\partial T/\partial x = 0$
- insulation condition for $T$ channel's bottom and ceiling surfaces: $\mathbf{n} \cdot \nabla T = 0$
- energy and mass continuity through the substrate exposed surface are enforced, except for liquid water for which this surface is impermeable: $\overline{T}_a = T_s$, $c_{vs} = c_{va}$, $\mathbf{n} \cdot \nabla c_{ls} = 0$
- no-slip and a standard logarithmic "wall functions" for velocity and the turbulence parameters, at any solid surface, are specified depending on the CFD code employed

The integration of the governing Equations is performed along the same lines already reported in the previous example of Sect. 6.2, in particular by identifying the most cost-effective grid, also reported in Fig. 6.19, that would allow for a adequate validation with the available experimental data.[7]

---

[6] Dry/basis humidity $X$ (kg/kg d.b.) and wet/basis humidity $U$ (kg/kg w.b.) values and concepts are often interchanged, with their implied meaning included in the dimensions. $X$ and $U$ are related by

$$U = \frac{X}{X + 1}$$

while the following conversion formulas with the molar concentrations hold:

$$c_l = \frac{1000 U \rho_s}{M} \qquad c_v = \frac{1000 \omega_a \rho_a}{(\omega_a + 1) M}$$

where the $T$-dependent density of humid air $\rho_a$ and the *specific humidity* $\omega_a$ (g/kg dry air) can be deduced by using a common *psychrometric chart*.

The specific humidity $\omega_a$ in the humid air has a parallel in the moist substrate with its *water activity* $a_w$.

[7] A complete account on the experimental validation can be found in the accompanying paper.

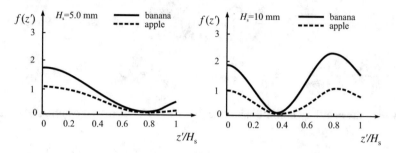

**Fig. 6.18** Distribution of $f$ along the normal coordinate $z$ for two common fresh foodstuffs and two values of thickness

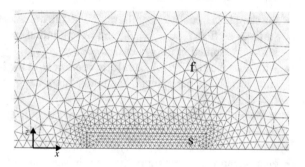

**Fig. 6.19** Close-up of the adopted grid, with notation of the fluid and solid phases. Reprinted from Journal of Food Engineering, Vol 95, Francesco Marra, Maria Valeria De Bonis, Gianpaolo Ruocco, Combined microwaves and convection heating: A conjugate approach, 31–39, Copyright (2010), with permission from Elsevier

Upon solution of the flow, temperature, and concentration fields, temperature and residual moisture maps for three main processing cases are shown in Fig. 6.20, in order to compare

(a)  pure convection drying
(b)  pure MW drying
(c)  MW-enhanced convection drying

The considered samples are disks of common fresh common potato of given thickness and volume $V$. Air process temperature is $T_p = 300$ K, and when present, air convection nominal velocity is $\mathbf{v}_p = 1.5$ m/s and nominal MW power is $e_0''' V = 250$ W. Process duration is the same, $\Delta\theta_p = 400$ s.

The classical external convective heating and related drying in Fig. 6.19a is commented first. A rather superficial heating is provided at sample's leading edge at left, corresponding to a quite ineffective drying as $X$ is rather uniform in the sample. The conjugate approach employed allows to couple the velocity field with temperature; therefore, the trailing edge is slightly colder and subject to a warmer flow wake.

Heating by pure MW is much faster and deeper as shown in Fig. 6.19b, where the temperature distribution is practically uniform, and so it is also for the moisture, subject to a much faster evaporation rate, driven by the buoyant thermal plume. $T$ and $X$ are much affected by the natural convection, which is correctly modeled, for the environment stagnant air is still colder than the processed substrate.

**Fig. 6.20** Quantitative temperature $T$ and residual humidity $X$ maps and contours for three main processing cases. The provided colormaps are consistent among the three figures, by using the same value ranges. *Source* see Fig. 6.19

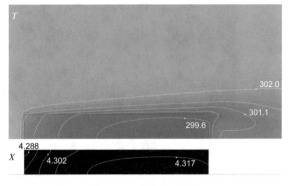

(a) pure convection drying

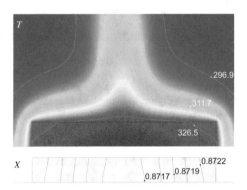

(b) pure MW drying

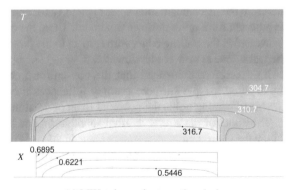

(c) MW-enhanced convection drying

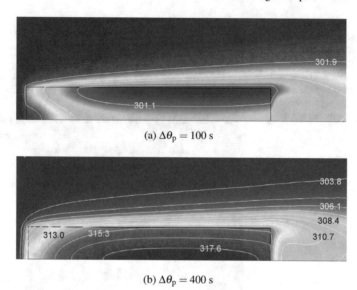

(a) $\Delta\theta_p = 100$ s

(b) $\Delta\theta_p = 400$ s

**Fig. 6.21** Quantitative temperature $T$ maps and contours for two different process durations in MW-enhanced convective drying. This time, for the sake of evidence the resulting thermal fields, the provided colormaps are not consistent between the subfigures. *Source* see Fig. 6.19

The combined treatment of MW-enhanced convection is finally shown in Fig. 6.19c. This process mode holds the favorable features of both mechanisms: the sample is treated faster that with sole MW, but on the leading edge a moisture excess still exist. The result is due to thermal mechanism dominance at the given process duration: therefore the mechanism intensities ($T_p$ and $\mathbf{v}_p$ for convection, and $e_0'''$ for MW) must be adjusted for a more uniform treatment. Then, the convective behavior (heat flux) is inverted at the trailing edge, with respect to the pure convection case in Fig. 6.19a: the wake zone is now colder downwind.

It is therefore evidenced that the non-uniform convective heat transfer, along the exposed sample surface, strongly influences the thermal treatment and that a combined heating holds the potential to an optimal mass transfer treatment.

The effect of the strong thermal source term can be inspected in Fig. 6.21, even within the simplified assumption of plane incident MWs. Here, the thermal evolution of the sample in the MW-enhanced convection drying, with the same operating parameters as the case in Fig. 6.19c, is explored, for two different process durations ($\Delta\theta_p = 100$ or 400 s). It is important to note that the relatively low source term employed here is justified so evidence the role of air convection in a fairly long duration. At the end of the first period, Fig. 6.21a, the sample is heated by about 8 K with an area, close to where the air impacts at first, characterized by a slight additional heating due to the convective heat transfer effect. As the sample continues to absorb MWs, Fig. 6.21b, the sample core undergoes to a further significant increase in its temperature, while the outer layers, whose temperature now exceeds that of the working air, release heat to the air flow.

The non-uniformity of the external convection is well marked by the temperature differences detected at the leading and trailing edges: the former (on which the fluid impacts) exhibits a higher cooling, this effect persisting and increasing as the process evolves in time. The external convection works, therefore, in two different ways: mitigating of a possible surface overheating in the upwind region, while almost insulating the downwind region due to the recirculation pattern. This must be taken into consideration because in an industrial process, when many samples are placed in a tunnel with continuous combined effect, the distance between the products can become a key parameter for obtaining uniform heating conditions.

## 6.4  Formation and Distribution of Harmful Compounds in Foodstuff Due to Applied Heat

The third and last important feature of multiphysics modeling is the ability to include biochemical and biological evolution into the physical framework of transfer phenomena. In the present case, we want to monitor undesired biochemical compounds that may form in substrates under strong heat exchange during time. The formation of some harmful constituents can be virtualized before the actual process, in order to execute a number of "what if" scenarios on initial biochemical budget and process operation so to minimize their occurrence and ultimately allow one to inspect the intimate working of this process.

Acrylamide (AA) is one of many such compounds that can be developed during heating of some foodstuff above a certain threshold. As AA is classified by IARC (International Agency for Research on Cancer) as a probable human carcinogen, the knowledge of critical processing variables leading to its formation is needed to ensure food safety requirements. AA forms in carbohydrate-rich substrates by means of the *Maillard reaction*, especially with a certain proportion between amino acids and sugars.[8]

In this case, each mass transport mechanism is strongly influenced by other ones: AA formation during starch-rich substrates affected the transformation of chemical species and the transfer of heat that need to be solved simultaneously in both solid and fluid phases. These phenomena are also strongly coupled through evaporation of the inherent substrate moisture during treatment. Therefore, AA formation is devised to a model fresh potato cube (phase s), immersed in an electrically heated oil bath (fluid phase f, Fig. 6.22). To model a realistic frying process, the sample is almost

---

[8] As reviewed for example by Carrieri, G., De Bonis, M.V., Pacella, C., Pucciarelli, A., Ruocco, G. Marra, F., De Bonis, M.V., Ruocco, G.: Modeling and validation of local acrylamide formation in a model food during frying. Journal of Food Engineering (2009) https://doi.org/10.1016/j.jfoodeng. 2009.04.017 and Carrieri, Anese, M., G., Quarta, B., De Bonis, M.V., Ruocco, G.: Evaluation of acrylamide formation in potatoes during deep-frying: The effect of operation and configuration. Journal of Food Engineering (2010) https://doi.org/10.1016/j.jfoodeng.2009.12.011.

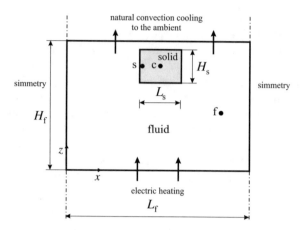

**Fig. 6.22** A schematic view of the subject two-dimensional container, with the associated lengths and coordinates, and indication of the core c, surface s, and fluid f thermocouple locations, needed for proper thermal validation. Reprinted from Journal of Food Engineering, Vol 98, Gabriella Carrieri, Monica Anese, Barbara Quarta, Maria Valeria De Bonis, Gianpaolo Ruocco, Evaluation of acrylamide formation in potatoes during deep-frying: The effect of operation and configuration, 141–149, Copyright (2010), with permission from Elsevier

surfacing in the bath. Phase s evolves during process in that its temperature and species (liquid water, water vapor, AA reactants, and AA proper) content changes locally and in time. During treatment, heat is transferred by natural convection from phase f to phase s surface and by conduction from the surface toward phase s interior where moisture vaporization also occurs.

So, convection, conduction, and vapor transfer are all responsible for the temperature increase in the sample, specially in zones that are closest to the working fluid, therefore favoring the creation of AA. Within the sample itself, instead, the temperature increase is slower due to the latent cooling offered by the inherent liquid water, which abounds in the sample fresh flesh. Following evaporation, this water diffuses through the substrate and surface the sample, then buoys and boils in the oil. The mechanism is partially impeded by the counter-diffusion of the oil, at least in the outer skin, and even prevented eventually when a thick and rigid crust is formed. These phenomena intensify with the increase of oil temperature and substrate thinning. With respect to the previous example in Sect. 6.3, the case at hand does not feature the intense source term but requires the modeling or fairly complex biochemistry. The following *assumptions* are considered:

- the oil is subject to laminar flow due to natural convection and is incompressible with constant properties (except for density)
- the bath extension is such that the system is assumed as infinite in both transversal and longitudinal directions, retaining its real height only (in order to monitor for the correct flow onset)

- no-slip condition is assumed at the processor bottom, and slip condition is assumed at the bath open surface
- perfect conductivity is assumed at the container bottom wall
- for the sake of simplicity, the food properties are temperature- and moisture-independent, and the viscous heat dissipation is neglected
- no crust modification nor geometry (shrinkage or swelling) effect is considered for the solid phase
- no AA nor liquid water is allowed to flow into oil, and no vapor is allowed to flow into the external ambient (the bath surface is isolated for the mass transfer); once the liquid water is converted into water vapor, it can diffuse in the solid phase and be transported in the fluid phase
- three AA reactants are considered only (fructose, glucose, and sucrose), their contribution to AA formation being merely additive
- the diffusive transport of AA and its reactants is negligible, as well as their kinetics in the fluid phase

With reference to the previous statements, the same governing Equations of the example in Sect. 6.3 for energy and momentum transfer can be adopted, with nothing new to be added. Instead, the governing Equations for mass transfer, written in terms of *local concentration of species c* (mg/g), are as follows:

*Mass transport, in the solid phase*  The transport of sugars is governed by

$$\frac{\partial c_R}{\partial \theta} = -K_{for} c_R \tag{6.6}$$

where $K_{for}$ (1/s) is a *rate of formation*. Then, the transport of AA is governed by

$$\frac{\partial c_{AA}}{\partial \theta} = K_{for} c_R - K_{eli} c_{AA} \tag{6.7}$$

where $K_{eli}$ (1/s) is a *rate of elimination*.

*Mass transport, in the fluid phase*  In this case, the governing Equations modify into:

$$\frac{\partial c_R}{\partial \theta} = -\mathbf{w} \cdot \nabla c_R$$

$$\frac{\partial c_{AA}}{\partial \theta} = -\mathbf{w} \cdot \nabla c_{AA}$$

The mechanisms of AA formation and elimination in this substrate can be described with a first-order kinetics: AA content in food results from two consecutive reactions, transforming from asparagine, *Asn*, and reducing sugars, *S*, to AA-protein complex/AA degradation product, *D*:

$$Asn + S \xrightarrow{K_{for}} AA \xrightarrow{K_{eli}} D$$

Thus, $K_{for}$ and $K_{eli}$ in Eqs. (6.6, 6.7) follow a modification of a first-order Arrhenius general form:

$$K = K_{ref}^n \exp\left[\frac{E_a}{R}\left(\frac{1}{T_{ref}} - \frac{1}{T}\right)\right]$$

with the exponent $n$ linearly dependent on the operating temperature $T_o$

With reference to Fig. 6.22, the *initial and boundary conditions* are as follows:

- the fluid phase is initially quiescent at $T_o$, while the solid phase is at the ambient temperature $T_\infty$, with a given initial moisture and reactant concentrations (raw sample). No vapor nor AA is initially assumed in the system
- slip condition with empirically imposed heat flux and insulation conditions for all pertinent species/components $i$ at the upper (horizontal) fluid surface, in contact with the quiescent ambient air at $T_\infty$:

$$\mathbf{n}\cdot\mathbf{u} = 0, \quad \mathbf{t}\cdot\left(-\nabla p + \mu_f\nabla^2\mathbf{u}\right)\mathbf{n} = 0$$

$$\mathbf{n}\cdot k\nabla T = -h_\infty\left(T - T_\infty\right)$$

$$\mathbf{n}\cdot\left(-D_i\nabla c_i + c_i\mathbf{u}\right) = 0$$

- no-slip condition with imposed temperature, and insulation conditions for all pertinent species/components $i$ at the bottom surface, as the operating temperature is regulated by the electric heater's thermostat:

$$\mathbf{u} = 0$$

$$T = T_o$$

$$\mathbf{n}\cdot\left(-D_i\nabla c_i + c_i\mathbf{u}\right) = 0$$

- symmetric condition at the vertical sides, where all considerations made earlier in the governing equations apply:

$$\mathbf{n}\cdot\mathbf{u} = 0, \quad \mathbf{t}\cdot\left(-\nabla p + \mu_f\nabla^2\mathbf{u}\right)\mathbf{n} = 0$$

$$\mathbf{n}\cdot\nabla T = 0$$

$$\mathbf{n}\cdot\left(-D_i\nabla c_i + c_i\mathbf{u}\right) = 0$$

- no-slip, energy, and mass continuity for vapor only, through the phase separation surface, as earlier implied:

$$\mathbf{u} = 0$$

$$T_f = T_s$$

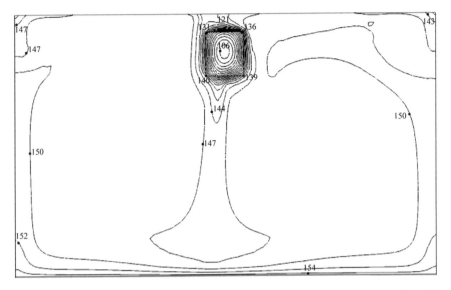

**Fig. 6.23** Temperature distribution, in the bath and in the sample, for deep-frying at 155 °C, at the end of treatment. *Source* see Fig. 6.22

$$\mathbf{n} \cdot \nabla c_i = 0 \quad \text{except for vapor}, \quad c_{fv} = c_{sv}$$

The integration of the governing equations is performed along the same lines already reported in the previous example of Sect. 6.3. The simulations were performed with the following data: $H_f = 0.059$ m, $L_f = 0.110$ m (the electric heater base being actually $0.26 \times 0.16$ m), $H_s = L_s = 0.010$ m, $T_o = 155$ or $180$ °C, $T_\infty = 19$ °C. Upon solution of the flow, temperature, and concentration fields, temperature and residual moisture maps can be presented for a variety of processing cases.

The contour plot of temperature distribution is shown in Fig. 6.23. The toroidal flow field that is formed during processing, due to natural convection onset, is slightly altered by the floating potato sample: the heated oil is subject to strong buoyancy near the lateral wall, and then, it is limited to a descending flow (downward plume) at the container center and starting from the immersed sample. The thermal field in Fig. 6.23 reflects such pattern. The sample temperature field is also non-uniform, the coolest value being found at its center, due to the residual evaporation cooling. The hottest sample temperature is reached instead at its lower corners at about 137 °C.

Some three-dimensional representations with added surface lighting for the AA distribution are then reported in Fig. 6.24. These representations emphasize on the surface concentrations, as the internal one is hidden in the protrusion, being much lower than the surface one. AA concentration is strongly related to the temperature

**(a)**                                                **(b)**

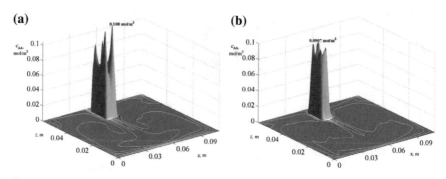

**Fig. 6.24** Three-dimensional map of local AA concentration (at the end of treatment), for deep-frying at 155 °C (left) or 180 °C (right), with indications of the highest AA values. *Source* see Fig. 6.22

field, as expected; therefore, $c_{AA}$ is non-uniform with its highest values being detected in correspondence of the lower (Fig. 6.24a) or upper corners (Fig. 6.24b). For longer process duration in Fig. 6.24b, at right, the maximum value of $c_{AA}$ is 10% less than the value found for the shorter process and lower temperature in Fig. 6.24a. Nonetheless, the concentration in Fig. 6.24b appears somewhat higher and more uniform along the surface, with less pronounced peaks. Therefore, it is confirmed that the AA kinetics is sensitive to both operating temperature and process duration, with a much stronger, nonlinear dependence with the former. In fact, a 1.fivefold AA formation increment corresponds to a modest 5% temperature increase (in K), while the AA maximum concentration is reduced almost linearly (a twofold AA decrement with a threefold duration shortage).

A first sensitivity case has been run by computing the average AA concentrations and the temperatures for three sample geometries, having the same volume but different rectangular aspect ratios, namely 1:1 and 4:0.25. These results in form of temperature maps and the flow fields are reported in Fig. 6.25. With the 1:1 aspect ratio (Fig. 6.25a), a close-up of what observed earlier in Fig. 6.23 is provided, by combining arrow velocity and isotherms: the flow field is nicely symmetrical below the cooler sample, forming the usual strong downward plume. With the other limiting aspect ratio of 4:0.25 (Fig. 6.25b), it is seen that the flow field is heavily altered instead, due to a lateral recirculation pattern, which in turn forms a thermal boundary layer on the sample upper side. This pronounced region works as a cooling heat pipe as the oil is driven and confined along the entire sample length while being cooled off by the bath–environment interface and the sample itself. Therefore in this region, the oil temperature is comparatively lower than elsewhere in the bath, which justifies a lower rate of AA formation. This effect, found also for at the higher operating temperature, makes the AA for the slice-shaped sample increasing moderately only, or even becoming constant for frying at 180 °C. This results confirms that the AA forming mechanism is driven, on a local base, by the availability of surfaces exposed to a favorable thermal configuration (Fig. 6.26).

**(a)**

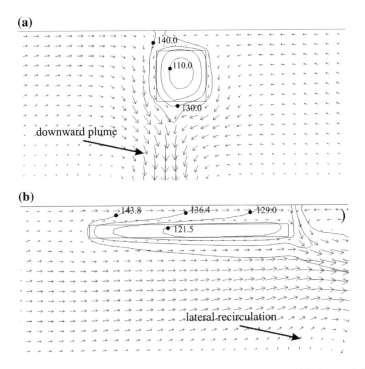

**(b)**

**Fig. 6.25** Flow fields and temperature maps subject to deep-frying at 155 °C, for **a** a 1:1 sample (maximum velocity $6.0 \times 10^{-3}$ m/s), and **b** a 4:0.25 sample (maximum velocity $4.7 \times 10^{-3}$ m/s). *Source* see Fig. 6.22

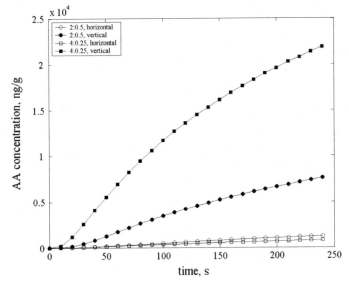

**Fig. 6.26** Evolution of computed average AA concentrations, subject to deep-frying at 155 °C, for two sample geometries and orientations. *Source* see Fig. 6.22

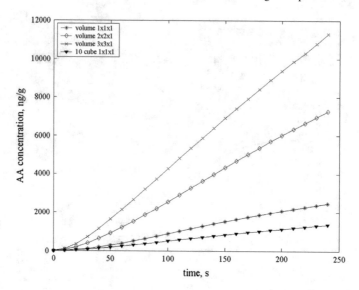

**Fig. 6.27** Evolution of computed average AA concentrations depending on sample mass, for deep-frying at 155 °C. *Source* see Fig. 6.22

## 6.5  Transport in Biostructures: From Microbial Spread to Tumor Growth Modeling

This last modeling example deals tackles with a couple of important topics in the biosciences (Fig. 6.27).

One interesting ability of transport phenomena framework is to assist the predictive microbiology in foodstuffs by means of dedicated modeling. A considerable hurdle in this case is the description of an inhomogeneous bacterial environment in a realistic substrate, including all of its structure features. Fresh leafy vegetables such as "iceberg" lettuce are common examples of structured substrates (Fig. 6.28). Lettuce cultivars are not only the most popular vegetable eaten raw, but also involved in severe pathogen outbreaks worldwide, such as for *E. coli*. Transport phenomena modeling can encompass realistic bacteria spreading in complex structures under a variety of operating conditions.

Another interesting scenario concerns the use of biological components for Moon and Mars exploration. In fact, interplanetary launch costs prohibit the establishment of permanent manned outposts for which most consumables would be sent from Earth. This issue could be alleviated producing part or all of these consumables on-site. To this end, the use of specific cyanobacteria (blue-green algae) is envisaged for the production of food, fuel, and oxygen,[9] but also for the inclusion

---

[9]Verseux, C. et al.: Sustainable Life Support on Mars - The Potential Roles of Cyanobacteria. International Journal of Astrobiology (2016) https://doi.org/10.1017/S147355041500021X.

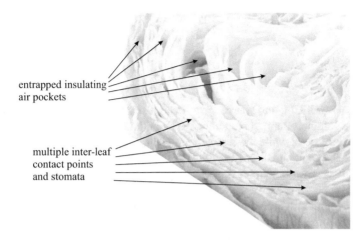

entrapped insulating
air pockets

multiple inter-leaf
contact points
and stomata

**Fig. 6.28**  Close-up of an actual iceberg lettuce, with indication of specific structural loci. Reprinted from Journal of Food Engineering, Vol 190, Maria Valeria De Bonis, Gianpaolo Ruocco, A heat and mass transfer perspective of microbial behavior modeling in a structured vegetable food, 72–79, Copyright (2016), with permission from Elsevier

in life-supporting structures, that would inherently contribute to harmful radiation shielding and absorption.

Using energy and biomass conservation complemented by kinetics notations, we can lead to description of actual bacterial behavior in all such substrates. A proof-of-concept model has been performed on a common produce item (Fig. 6.28).[10] The main features of the model are the following:

1. *Bacterial kinetics*, and bacterial cell population at any point in the substrate
2. *Bacterial migration*, which dictates biomass diffusion through the substrate
3. *Structural features of the colonized substrate*, in association with bacterial migration. In the case at hand, considering an ensemble of leaf shells of given thickness (one inside the other), we end up in many random inter-leaf contact points and "entrapped" air pockets or interspaces (Fig. 6.28). Consequently, the bacterial population is expected to spread through the produce by exploiting the former, while its migration would be hindered due to the latter
4. *Thermal regime variability*, as any combination of time/temperature can be speculated to come up with indications of bacterial behavior and conditions of the structure: in the case at hand, the safety from harvest to consumer.

The governing equations in vector form and homogeneous properties for the *heat transfer* and *E. coli biomass kinetics and transfer* are therefore applied to a quasi-spherical lettuce model of given average radius. With the spherical symmetry exploited, the PDE system includes the so-called Bioheat Eq. (6.8) for the distributions of temperature $T$, and bacterial cell population $x$ species balance Eq. (6.9):

[10]De Bonis, M.V., Ruocco, G.: A Heat and Mass Transfer Perspective of Microbial Behavior Modeling in a Structured Vegetable Food. Journal of Food Engineering (2016) https://doi.org/10. 1016/j.jfoodeng.2016.06.015.

$$\rho c \frac{\partial T}{\partial \theta} = k \nabla^2 T \tag{6.8}$$

$$\frac{\partial x}{\partial \theta} = D \nabla^2 x + S \tag{6.9}$$

Here, $[x] = \log CFU/g$, and source term $S$ can be cast based on a logistic function, following a common Baranyi paradigm.[11] The initial temperature is uniformly set at $T_i$ in the lettuce item, while the temperature is uniformly set on the external surface depending on the applied thermal regime (cool storage, distribution, sale). Moreover, the initial bacterial population is set at a given function within the lettuce item, depending on possible extrinsic contamination and/or internalization of pathogen.

Insulation is applied at the external surface, as cell bacterial population cannot spread into the environment. Continuity is applied for bacterial migration through each inter-leaf contact point, while insulation is applied again for bacterial migration at each insulating inter-leaf surface, and as a result, consecutive leaves are insulated everywhere except at internal contact points.

As a common scenario, a value of 4.8 $\log CFU/g$ is speculated on the external leaf of the sample, with an given effective bacterial diffusivity, to simulate an external contamination. With the exception of the first leaf, the vegetable is initially uninfected. The initial temperature is uniform and equal to the operating one. Standard properties are applied.

Figure 6.29 shows the bacterial cell population distribution $x$ at varying external temperatures ($T_e = 10$ or 25 °C) after a storage duration of $\Delta\theta = 10$ h. It is seen that a different number of leaves are consecutively populated in the allotted time, the higher $T_e$ allowing for a deeper contamination from the external leaf. In order to monitor the contamination intensity and depth, $x_{avg}$ was computed through a volume integration extended to the leaf denoted by its index L$i$. When $T_e = 10$ °C (Fig. 6.29, left), the average cell population of last contaminated leaf L2 is $x_{avg,L2}\big|_{T_e=10\ °C} = 0.20\ \log CFU/g$. In comparison, when $T_e = 25$ °C (Fig. 6.29, right), a similar level of local contamination is found in the last contaminated leaf L8 ($x_{avg,L8}\big|_{T_e=25\ °C} = 0.27\ \log CFU/g$).

In a similar way, a second, even more striking task in the bioscience modeling, is the understanding of transport phenomena involved in tumor formation and progression. Just as well in this case, the scrutiny of the biomass in space and time has to take into account the inherent *tissue* anisotropy and inhomogeneity, while the perturbations are due to mutations, drug effects, or external stimuli such as growth factors. In this case, it is obvious that every modeling results and the robustness of sensitivity analysis should be validated with the available clinical data.

Computational advances make it now feasible to investigate tumor proliferation via an integrative approach that includes clinical examinations, engineering modeling, and *in-silico* simulations. Such a modeling approach can lead to a novel strategy of personalized medicine by generating detailed virtualized predictions for the ther-

---

[11] Baranyi, J. et al.: Brochothrix Thermosphacta at Changing Temperature. International Journal of Food Microbiology (1995) https://doi.org/10.1016/0168-1605(94)00154-X.

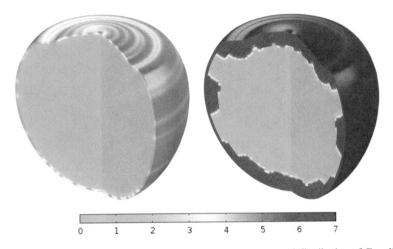

**Fig. 6.29** External contamination at constant temperatures: computed distribution of *E. coli* cell population $x$ (logCFU/g) for two values of $T_e$ and for $D = 1 \times 10^{-10}$ m$^2$/s, after $\Delta\theta = 10$ h and when $x_i(R - \delta \le r \le R) = 4.8$ logCFU/g. Left: $T_e = 10$ °C; right: $T_e = 25$ °C. *Source* see Fig. 6.28

apy of cancer and could be integrated in routine diagnostics in oncology, to realize scenarios of augmented reality.

The main features of the model are the followings:[12]

1. *Tissue microscopic topology and typology*, describing type, density, and cell fraction of the given tissue in a given human organ (such as the one depicted in Fig. 6.30)
2. *Interstitial pressure*, which depends on the dynamic tissue topology at any point in the organ
3. *Cell activity*, depending on local metabolic activity which is increased for tumor cells
4. *Nutrients, metabolic products, growth factor, and drug concentrations*, whose governing Equations are intertwined to simulate the real behavior of tumors

As modeling tackles more susceptible problems, tissue property perturbations must be taken into account to modify the known parameters of the model. Multiobject optimization can also be employed to identify difficult-to-determine or unknown parameters. Robust Computer-Aided Engineering techniques need to be enforced to superimpose modeling to diagnostic imaging.

Tumor growth and angiogenesis can be resolved by the following comprehensive model applied to the simplified geometry depicted in Fig. 6.30. The PDE system includes again forms of the continuity Eq. (6.9) for a variety of chemical species: concentration of oxygen $c_{O_2}$ (nutrient), carbon dioxide $c_{CO_2}$ (metabolites), tumor

---

[12]Tang, L. et al.: Computational Modeling of 3-D Tumor Growth and Angiogenesis for Chemotherapy Evaluation. PloS one (2014) https://doi.org/10.1371/journal.pone.0083962.

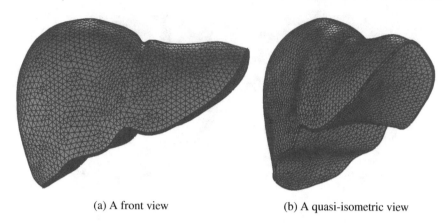

(a) A front view                              (b) A quasi-isometric view

**Fig. 6.30** An discretized 3-D geometry of a human liver. More realistic renderings are possible when the available diagnostic imaging are acquired by the CAE bundle

angiogenesis factor $c_{TAF}$, and specific drug $c_d$. The main lines of the model are as follows:

1. recognition of the diffusivities $D_i$ for each of the variables, in each of the tissues involved
2. geometry calculations (based on cell and vascular density around each tissue element considered, in the mathematical sense) lead to the distribution of the total tumor interstitial pressure $p$, due to cell-induced pressure and vascular perfusion-induced pressure
3. cell cycle is determined by a local cell activity parameter $A$, which is defined depending on the local budget of nutrients and metabolites
4. from the distribution of pressure, the interstitial fluid velocity is determined via Darcy's law
5. fluid velocity $\mathbf{u}$ determines the increment of diffusion term in each of the four governing equations by adding a suitable $D_{\mathbf{u}}$:

   a. oxygen concentration $c_{O_2}$. As glucose, amino acids, fatty acids, vitamins, and micronutrients are hydrophilic and do not easily permeate across cell's plasma membrane, for the sake of simplicity this is the representative cell nutrient for DNA, RNA, and proteins synthesis and affects tumor cell metabolism, angiogenesis, growth, and metastasis:

$$\frac{\partial c_{O_2}}{\partial t} = \left( D_{O_2} + D_{\mathbf{u}} \right) \nabla^2 c_{O_2} + S_{O_2}$$

   Source term $S_{O_2}$ consists in:
   • a positive term due to the oxygen released by blood cells supplied locally by blood vessel, depending on pressure difference across vessel wall and vessel size

- a negative term due to the oxygen uptake by cells, proportional to the local $A$ value

b. carbon dioxide concentration $c_{CO_2}$. As metabolic wastes accumulation occurs during tumor growth process due to inefficient drainage, causing reduced cell activity and biosynthesis up to necrosis, for the sake of simplicity this is the representative cell waste metabolite and affects tumor growth:

$$\frac{\partial c_{CO_2}}{\partial t} = \left(D_{CO_2} + D_u\right) \nabla^2 c_{CO_2} + S_{CO_2}$$

Source term $S_{CO_2}$ consists in:
- a positive term due to the carbon dioxide release rate, proportional to the local $A$ value
- a negative term due to the carbon dioxide uptake by blood cells supplied locally by blood vessels, depending on pressure difference across vessel wall and vessel size

c. Tumor Angiogenesis Factor (TAF) concentration $c_{TAF}$. In this preliminary model, this variable underlies a variety of biological capabilities, qualified as rational condition for neoplastic diseases. As tumor cells rapidly consume nutrients, hypoxia occurs in the center of avascular tumors once a given threshold is reached, causing TAF secretion to favor new vessels to sprout from existing ones toward hypoxia regions:

$$\frac{\partial c_{TAF}}{\partial t} = \left(D_{TAF} + D_u\right) \nabla^2 c_{TAF} + S_{TAF}$$

Source term $S_{TAF}$ consists in:
- a positive term due to the TAF release rate, proportional to the local $c_{O_2}$ value
- a negative term due to the TAF uptake by blood cells supplied locally by blood vessels, depending on pressure difference across vessel wall and vessel size

d. drug concentration $c_d$. Chemotherapy is an important element to attack tumors and metastases. The effect of anti-angiogenic drugs is to reach a local concentration above a given threshold in order to lower the TAF release rate or inducing direct apoptosis; while cytotoxic drugs above a given threshold induce DNA damage to tumor cell in order to prevent or limit cell replication

$$\frac{\partial c_d}{\partial t} = \left(D_d + D_u\right) \nabla^2 c_d + S_d$$

Source term $S_d$ consists in:
- term being positive or negative, due to the drug release and uptaken by blood cells supplied locally by blood vessels, depending on pressure difference across vessel wall and vessel size
- a negative term due to the drug natural decay

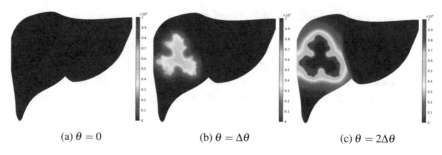

(a) $\theta = 0$      (b) $\theta = \Delta\theta$      (c) $\theta = 2\Delta\theta$

**Fig. 6.31**  Initial tumor progression

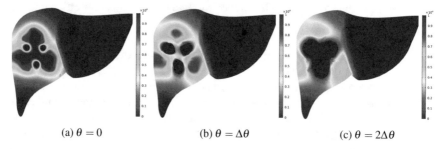

(a) $\theta = 0$      (b) $\theta = \Delta\theta$      (c) $\theta = 2\Delta\theta$

**Fig. 6.32**  Tumor progression during successful therapy

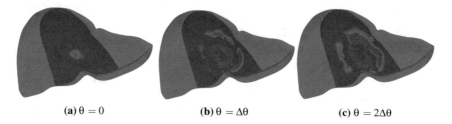

(a) $\theta = 0$      (b) $\theta = \Delta\theta$      (c) $\theta = 2\Delta\theta$

**Fig. 6.33**  A tumor occurrence and localized therapy effect

A lateral cross section of a geometry similar to the one depicted in Fig. 6.30 was considered. Cancer cells migration and the tumor proliferation were initiated. In Fig. 6.31, the evolution of the tissue was detected at three times, with no therapy applied so that cancer cells freely migrate in the affected part. In Fig. 6.32, a drug therapy was applied in three distinct areas of the organ, so that the combined and intertwined (nonlinear) effect of variable concentrations shows that the tumor proliferation decreases over the time, the cancer cells migration being arrested, and new tissue generated.

In another computational trial of Fig. 6.33, a developing tumor was initially visualized in the liver of Fig. 6.30, while a successful drug therapy was applied which restored somehow the tissues, specially in the vicinity of the initial region of interest, while a wide proliferation is developing in an untreated lateral region.

## Reference

1. Gerald, C.F., Wheatley, P.O.: Applied Numerical Analysis. Addison-Wesley, Reading (1984)

# Correction to: Introduction to Transport Phenomena Modeling

**Correction to:**
**G. Ruocco, *Introduction to Transport Phenomena Modeling*,**
**https://doi.org/10.1007/978-3-319-66822-2**

In the original version of the book, the belated corrections from author for Chaps. 2–6 should be incorporated. The correction book has been updated with the changes.

The updated version of these chapters can be found at

https://doi.org/10.1007/978-3-319-66822-2_2

https://doi.org/10.1007/978-3-319-66822-2_3

https://doi.org/10.1007/978-3-319-66822-2_4

https://doi.org/10.1007/978-3-319-66822-2_5

https://doi.org/10.1007/978-3-319-66822-2_6

# Index

© Springer International Publishing AG 2018
G. Ruocco, *Introduction to Transport Phenomena Modeling*,
https://doi.org/10.1007/978-3-319-66822-2